Studies in

Subjective Probability

STUDIES IN
SUBJECTIVE PROBABILITY

edited by

HENRY E. KYBURG, JR.
University of Rochester

HOWARD E. SMOKLER
University of Colorado

ROBERT E. KRIEGER PUBLISHING COMPANY
HUNTINGTON, NEW YORK
1980

Original edition 1964

Second edition 1980

Printed and Published by
ROBERT E. KRIEGER PUBLISHING COMPANY, INC.
645 NEW YORK AVENUE
HUNTINGTON, NEW YORK 11743

Copyright © 1964 (original material) by
John Wiley & Sons, Inc.
Copyright © 1980 (new material) by
ROBERT E. KRIEGER PUBLISHING CO., INC.

Printed in the United States of America

Library of Congress Cataloging in Publication Data

Kyburg, Henry Ely, 1928–   ed.
  Studies in subjective probability.

  Includes bibliographical references and index.
  1. Probabilities—Addresses, essay, lectures. 2. Logic—Addresses, essays, lectures. I. Smokler, Howard Edward, 1928– joint ed. II. Title.
BC141.K92   1979   160   79-16294
ISBN 0-88275-296-0

PREFACE

In the fifteen years since the first edition of *Studies in Subjective Probability* appeared, the point of view represented by de Finetti, Ramsey, and Savage has become better known not only in philosophy and statistics where it originated, but in schools of business management, in economics and formal political science, and in psychology, both as a theory to test and a method for testing theories. This hardly means that the view has swept the intellectual scene by storm. Philosophically it has always been highly controversial, and there are some indications that the tide of the controversy is beginning to run against it. In statistics and decision theory the elegance and simplicity of the basic ideas are mitigated to some extent by the complexity of the mathematics into which one is often led in practice. In the social sciences the impact has been very small, but appears to be growing.

Since this point of view is no longer a curiosity, but a respectable minority view, the function of such a book of readings as this is no longer quite what it was. It seems less appropriate to devote sections to the historical background and early intimations of the theory, now that so many people have some familiarity with the basic ideas. On the other hand the fundamental papers of Ramsey and de Finetti are both of historical interest and valuable to this day as clear expositions of the basic point of view. We have therefore dropped the articles by Venn ("The Subjective Side of Probability") and Borel ("Apropos of a Treatise on Probability") as of merely historical interest. We have replaced one article by Savage with a more recent one, and have added in addition to this scholar's work one by another giant of the field, de Finetti. We have also added a paper by I. J. Good which suggests one of the several directions in which the idea of subjective probability can be explored and developed, and a paper by Richard Jeffrey demonstrating a more purely philosophical application of the theory.

Henry E. Kyburg, Jr.
Howard E. Smokler

CONTENTS

HENRY E. KYBURG, JR., AND
HOWARD E. SMOKLER

Introduction

"Probability" is a word that is used constantly in empirical science, in mathematics, in philosophy, and in a multitude of situations in everyday life. It is a word which has (together with such synonyms as "likelihood," "chance," etc.) become more and more pervasive in science since the simple, deterministic laws of the eighteenth century have been supplemented, and sometimes superseded, by laws of a statistical or probabilistic character. In view of its frequent usage in a wide variety of contexts, it is not surprising that the word "probability" has acquired a number of shades of meaning which are not clearly distinguished from one another. In ordinary discourse it does not matter that a word has several shades of meaning, provided that the particular shade intended is made clear enough by the context in which the word is used. But "probability" is also a technical word, occurring in science and mathematics; when it is used in a technical way, it should have a clear and definite meaning in the particular context in which it is used. The same is true of its philosophical uses. Several proposals have been made regarding this meaning; the subjectivistic theory of probability provides one of them.

To begin with, some writers, such as Carnap[1] and Nagel,[2] allow that "probability" has two quite distinct sorts of meanings. For most of these writers, who include many who also hold a subjectivistic view of probability, one sense of the word concerns chance, or long run frequency, or propensity—that is, it is a sense which is empirical, objective, and quite independent of what anyone knows or believes. At the same time, there is another sense in which "probability" is used, in which it is taken to measure actual, or refined, or justified degrees of belief. Carnap distinguished these two senses as probability$_2$ and probability$_1$, respectively. Other writers have argued that all intelligible uses of "probability" can be explained by one interpretation. De Finetti is one of these; he regards the notion of chance, or objective probability, or Carnap's probability$_2$ as "metaphysical moonshine." At the other extreme, writers such as von Mises[3] and Reichenbach[4] regarded only probability$_2$, objective empirical probability, as of interest to either science or philosophy. A third unitary view is that of Kyburg,[5] according to which all probabilities represent logical relations. In what follows we shall leave to one side the complications which arise from the fact that a single writer may adopt two interpretations of probability simultaneously.

(1) Carnap [1962a], *The Logical Foundations of Probability*.
(2) Nagel [1937], *Principles of the Theory of Probability*.
(3) von Mises [1951], *Probability, Statistics, and Truth*.
(4) Reichenbach [1949], *The Theory of Probability*.
(5) Kyburg [1974], *The Logical Foundations of Statistical Inference*.

The basic ideas of the subjectivistic theory of probability can best be grasped by comparing it with some of the other proposals regarding the meaning of the term. One framework in which the term "probability" has a definite meaning is that of pure mathematics. It has been said (facetiously) that there is no problem about probability: it is simply a non-negative, additive set function, whose maximum value is unity. This is neat and unambiguous as far as it goes, but it is of no immediate help in explaining how probability is used in the particular sciences, or in such semi-technical contexts as "our empirical knowledge is at best probable"; it is of no help in explaining how probability and statistics are used by insurance companies, nor how it is that people should go about following Bishop Butler's maxim that one should take probability as a guide in life.

To take probability as simply a mathematical function of a certain sort is to take it as an undefined term in a formal system; but when we come to apply the formal system to the world, when we begin to talk about the probability of certain specific events or even of certain kinds of events, or when we take probability as a guide in life, we are driven to think more closely about the notion of probability itself. We must find some connection between this abstract entity which satisfies certain mathematical stipulations and the pragmatic content, the real meaning, of the important statements of scientific and social intercourse which contain the word "probability" or one of its synonyms.

There are essentially three types of connection that have been proposed: the empirical, the logical, and the subjective. The empirical, or frequentist, conception of probability appears in many guises and is very widely held today. As it was first formulated by Venn,[6] and later by von Mises[7] and Reichenbach,[8] this view identified probability with the limit of a relative frequency. To say that the probability that an A is a B is p, is simply to say that the limit of the relative frequency of B's among A's (as the number of observed A's is increased without bound) is p. There are difficulties connected with this view of probability (although it still has its champions, for example, Salmon[9]), and most statisticians and philosophers who hold an empirical view of probability prefer to treat it as a theoretical concept receiving its meaning from the rules for applying the theory. Braithwaite, for example, in chapters V, VI, and VII of this book *Scientific Explanation,*[10]

(6)   Venn [1886], *The Logic of Chance.*
(7)   von Mises [1951], *Probability, Statistics, and Truth.*
(8)   Reichenbach [1949], *The Theory of Probability.*
(9)   Salmon [1963], "On Vindicating Induction."
(10) Braithwaite [1953], *Scientific Explanation.*

provides a "theoretical model" conception, in which the probability that an *A* is a *B* is a *parameter* in a model, which is as a whole given empirical content by a "rule of rejection": if we draw samples of a certain determinate character (itself determined by a logical analysis of the model), then we reject the theoretical model as providing an adequate explanation of our observations of *A*'s and *B*'s. It is fair to say that most statisticians today hold views which, while not so formal or explicit as Braithwaite's, are not essentially different from his. Neyman[11] and the British-American school of statisticians, like Sir Ronald A. Fisher's[12] "likelihood" school, are concerned primarily with the formulation of desirable rules for the acceptance and rejection of statistical hypotheses on the basis of observed evidence; for these people a probability statement can only be a statistical hypothesis or a consequence drawn from a statistical hypothesis.

The important point is that a probability statement is taken as making an assertion about the world. It may be right or wrong—and it is generally held that we never really know with certainty which it is—but it is a statement, like a statement about lengths or weights, which is either true or false, and for which the evidence is chiefly observational. In order to find out whether or not a probability statement is true, we must make an empirical investigation, and usually this will be a non-terminating investigation of the sort whose results are said (in a non-empirical sense) to be only "probable."

The alternative to this view is to deny that probability statements are *empirical* statements at all; the extreme version of this alternative is to take probability as representing a logical relation between a proposition and a body of knowledge, between *one* statement and another statement (or a set of statements) representing evidence. Such a view was first formulated explicitly by Keynes[13] and has been defended by Carnap,[14] Hintikka,[15] and Kyburg,[16] among others. The essential characteristic of this view is this: given a statement, and given a set of statements constituting evidence or a body of knowledge, there is one and only one degree of probability which the statement may have, relative to the given evidence. A probability statement is *logically* true if it is true at all; otherwise it is *logically* false. Probability statements are purely formal in the same way that arithmetical statements are purely formal. Given a statement *S* and a body of evidence *E*, there is one and

(11) Neyman [1952], *Lectures and Conferences on Mathematical Statistics*.
(12) Fisher [1956], *Statistical Methods and Scientific Inference*.
(13) Keynes [1921], *A Treatise on Probability*.
(14) Carnap [1962], *The Logical Foundations of Probability*.
(15) Hintikka [1965], "On a Combined System of Inductive Logic."
(16) Kyburg [1974], *The Logical Foundations of Statistical Inference*.

only one real number p such that it may correctly be said that the probability of S relative to E is p. [17]

Carnap, who for many years was a major advocate of this position, introduced a distinction between pure and applied inductive logic. [18] Applied inductive logic is a discipline which arises in the context of human action. Superimposed upon a descriptive theory of decision-making is a normative theory of the same domain. The normative theory specifies the conditions for *rational* action. As belief is a necessary component of decision-making, so rational belief is a necessary condition of rational decision-making. Certain conditions placed upon bets (acts) insure that the odds given on the occurrence of an event (or set of events) provide a probability measure on the set of events. Further conditions placed upon bets insure that the probability measure never assigns the value 1 or 0 to the probability of an event unless the event is the necessary or impossible event. Carnap calls the measure of belief in H for a person X at a time T, that person's *credence function,* $Cr_{X,T}(H)$, for short $Cr_T(H)$. He argues further that rational change in belief over time should be a function only of the credence function at the beginning of that time and of the evidence that might be accumulated during the period after the initial assignment of belief. In other words, he postulates a permanent disposition to believe, which he calls a *credibility function, Cred* (H). For this function (but not for the credence function) Carnap postulated axioms of symmetry. These state that the credibility of two propositions, H and H', are equal if the only difference between them is that the individual name in H is replaced by another individual name. In addition to the axiom of symmetry, Carnap proposes as axioms the standard axioms of probability and one or two others, like the axiom of regularity and the axiom of convergence.

These axioms or principles of rational belief are stated in quasi-psychological terms. A pure inductive logic is a set of axioms corresponding to the axioms or principles of rational belief but stated in a purely logical way. Carnap has argued that the basis for these axioms is intuition. [19]

Carnap always cherished the idea that the axioms of inductive logic would rule out all but one acceptable probability function, although he never found a set of intuitively acceptable axioms that yielded this result.

(17) In Kyburg's system (*The Logical Foundations of Statistical Inference*), which also provides a logical interpretation of probability, a *pair* of fractions (p, q) plays the role of the single real number p, but the principle is the same; Kyburg [1974].

(18) Carnap [1971b], "A Basic System of Inductive Logic."

(19) Carnap [1968], "Inductive Logic and Inductive Intuition."

The subjective or personalistic view is to be distinguished from the logical view precisely by its denial of the assertion that only one probability or credibility function is rationally acceptable. On the subjectivistic view, probability represents a relation between a statement and a body of evidence, but it is not a purely logical relation. It is a quasi-logical relation and the numerical value attached to it represents a degree of belief. On the subjectivistic view, this value is *not* uniquely determined: a given statement may have any probability between 0 and 1, on given evidence, according to the inclination of the person whose degree of belief that probability represents. Of course, in the case in which the evidence logically entails the statement in question or entails its denial, the criteria of ordinary deductive logic are applicable.

Probability is not, however, merely psychological, and the subjectivistic theory of probability is not an empirical psychological theory of degrees of belief. There has been some confusion of this point which should, however, be cleared up, once and for all, by de Finetti's new footnote on page 71. The source of the confusion has been research which has been performed with a view to finding out whether people's degrees of belief are related in the ways they should be related according to the theory.[20] What is being tested here is not the normative theory, but its empirical underpinning. The two are closely interlocked: we may both assess our canons of rationality by looking at the way rational people behave, and assess the rationality of people by comparing their behavior to that indicated by our canons of rationality.

The subjectivistic theory of probability is a logical theory in the sense that only certain combinations of degrees of belief in related propositions are admissible. If a person has a degree of belief p in a statement S, then he *should* have a degree of belief $1 - p$ in the denial of S. The proponents of the subjectivistic theory attempt to justify this "should." They do so by arguing the degrees of belief can be measured by betting ratios (the least odds at which one is willing to bet on the truth of a statement), and that if one's degree of belief in S and in the denial of S, for example, did not have the relation suggested, then it would be possible to have book made against one in such a way that one was bound to lose regardless of whether S was true or false.

For example, according to this theory, to say that the probability of S is 1/4 for someone is to say that he is willing to bet 3 to 1 against its truth. If he assigned the same probability of 1/4 to the denial of S, the book could consist of two bets: one at odds of 1 to 3 against S, one at odds of 1 to 3 against the denial of S. If the units are dollars, this insures the book of a profit of two dollars: i.e., the person is bound to lose in any event. It needn't

(20) Edwards [1960], "Measurement of Utility and Subjective Probability."
        Slovic, Fischoff, and Lichenstein [1977], "Behavioral Decision Theory."

be argued that these bets are not desirable. It is tautological to say that one does not wish to lose objects with positive utility under all possible conditions. Having a distribution of degrees of belief which obey the conventional rules of the calculus of probabilities is a logically necessary and sufficient condition of not having book made against one.[21] The possession of such a distribution of degrees of belief is called *coherence* by the subjectivists. The demand that one should be *coherent* in one's degree of belief is thus a logical demand. This demand is made by holders of logical conceptions of probability as well as by the subjectivists. On the subjectivistic theory, however, it is the *only* demand that is made—and even this demand is considerably attenuated in Koopman's system, in which not all degrees of belief are comparable.

In the subjectivistic view, probability represents the *degree of belief* that a given person has in a given statement on the basis of given evidence. (Some theorists prefer to speak of "events" rather than "statements"; the difference is unimportant here.) The person ought to be *consistent* in the strict logical sense: if the evidence logically entails E, then he should have the highest degree of belief in E; if the evidence entails the denial of E, he should have the lowest possible degree of belief in E. This is as far as deductive logic takes us. The subjectivistic theory of probability takes us one step further: the person's body of beliefs considered as a whole should be *coherent* as well as *consistent*. The theory—or, better, theories—are subjective or personalistic in the sense that a person can have any degree of belief whatever in any given statement on any evidence, provided only that his other degrees of belief have suitable values. It is in this sense that the subjectivistic theory is subjectivistic.

Perhaps the simplest way to characterize the subjectivistic conception of probability is to say (a) that any degree of belief in any statement is permissible, but (b) that there are restrictions placed on the distribution of degrees of belief among sets of related statements. In either of the alternative views of probability—the empirical and the logical—there is one and only one degree of probability that can be correctly assigned to an event of a certain sort in a given sequence, or to a statement relative to a given body of evidence.

There remains the question of the way in which the normative function is to be performed. Not all distributions of belief are admissible on the subjectivistic theory. Suppose a person finds that his distribution of degrees of belief is *incoherent*. By having his attention called to the lack of coherence, the person (desiring to be logical, or to attain what he desires) will no doubt

(21) For a demonstration of this fact, see the following articles cited in the bibliography: Kemeny [1955], Shimony [1955], Lehman [1955].

be led to remove that incoherence, but the way in which he removes it is entirely his own affair. Some of his opinions (as reflected in his assignments of probability) must be modified, but the theory does not dictate in any way at all which of these opinions should be modified or how they should be modified. Any particular opinion may be maintained by suitably modifying the remaining opinions. In this sense, too, the subjectivistic theory is subjectivistic.

There are a few key terms that crop up frequently in discussions of the subjectivistic theory. The most important of these are: "degree of belief," "coherence," and "exchangeable events." In the following paragraphs we will sketch some of the historical development of these notions, and indicate their current meaning.

The history of probability theory is intertwined with the notions of belief and opinion, as well as with games of chance and statistics. As Hacking has pointed out, in the late medieval and early renaissance period probability was a characteristic of opinion and not knowledge; the latter was, of course, certain.[22] If we take "belief" as a synonym for opinion then on the traditional view probability is a mark of belief and not of knowledge. Furthermore, belief is susceptible of distinction: some beliefs are better grounded than others. Locke expresses both these ideas in his *Essay Concerning Human Understanding*, Book IV, Chapter XV, "Of Probability":

2. Our knowledge, as has been shown, being very narrow, and we not happy enough to find certain truth in everything which we have occasion to consider; most of the propositions we think, reason, discourse—nay, act upon, are such as we cannot have undoubted knowledge of their truth: yet some of them border so near upon certainty, that we make no doubt at all about them; but assent to them as firmly, and act, according to that assent, as resolutely as if they were infallibly demonstrated, and that our knowledge of them was perfect and certain. But there being degrees herein, from the very neighbourhood of certainty and demonstration, quite down to improbability and unlikeness, even to the confines of impossibility; and also degrees of assent from full assurance and confidence, quite down to conjecture, doubt, and distrust: I shall come now, (having, as I think found out *the bounds of human knowledge and certainty,*) in the next place, to consider *the several degrees and grounds of probability, and assent or faith.*

3. Probability is likeliness to be true, the very notation of the word signifying such a proposition, for which there be arguments or proofs to make it pass, or be received for true. The entertainment the mind gives this sort of propositions is called *belief, assent,* or *opinion,* which

(22)　Hacking [1975], *The Emergence of Probability.*

is the admitting or receiving any proposition for true, upon arguments or proofs that are found to persuade us to receive it as true, without certain knowledge that it is so. And herein lies the difference between *probability* and *certainty, faith,* and *knowledge,* that in all the parts of knowledge there is intuition; each immediate idea, each step has its visible and certain connexion: in belief, not so. That which makes me believe, is something extraneous to the thing I believe; something not evidently joined on both sides to, and so not manifestly showing the agreement or disagreement of those ideas that are under consideration.[23]

In such a scheme probability is an aspect of ignorance. James Bernoulli, one of the prominent early workers in the mathematical theory of probability, viewed it this way, as did Laplace, another great contributor to the field.[24] In his work *Ars Conjectandi* (1713), Bernoulli points out that since we cannot generally know with certainty whether or not an event will occur, we can only have a "degree of confidence" in the truth of the proposition that asserts its occurrence. This "degree of confidence" is identified with the probability of an event. It depends on the knowledge that the individual has at his disposal. Therefore it can vary from individual to individual. The art of guessing (*Ars Conjectandi*) consists in estimating as precisely as possible the best value of probabilities.[25]

A more modern writer who explicitly defines probability in terms of "degree of belief" is Augustus De Morgan (1847). He has this to say:

There is no further use in drawing distinctions between the knowledge we have of our own existence, and that of two and two amounting to four. This absolute and infallible feeling we shall call *certainty.* We have lower grades of knowledge, which we usually call *degrees of belief,* but they are really degrees of knowledge. . . .

It may seem a strange thing to treat *knowledge* as a magnitude, in the same manner as length, or weight, or surface. This is what all writers do who treat of probability, and what all their readers have done, long before they ever saw a book on the subject. But it is not customary to make the statement so openly as I now do, and I consider that some justification of it is necessary.

By degree of probability we really mean, or ought to mean, degree of belief. . . . Probability then, refers to and implies belief, more or less, and belief is but another name for imperfect knowledge, or it may be, expresses the mind in a state of imperfect knowledge.[26]

(23) Locke [1965], *Essay Concerning Human Understanding.*
(24) Laplace [1952], *Philosophical Essay on Probabilities.*
(25) Bernoulli [1713], *Ars Conjectandi.*
(26) De Morgan [1847], *Formal Logic,* pp. 171-173.

De Morgan thus states (1) that probability is identified as a "degree of belief," and (2) that such a degree of belief is a measurable quantity, as is theoretically certain knowledge.

De Morgan explicity recognizes that the dispositional property of a person

(1) *a* [believes that ϕ]

can be quantified and that therefore belief (or knowledge) is a measurable quantity. This is in direct analogy to the way that mass or distance is a quantity. So the question becomes: do the conditions for proper measurement of belief occur here? Many early frequentists denied this possibility, but others did not give up hope.

John Maynard Keynes, in his book *A Treatise on Probability* (1921), introduced several important ideas into the foundations of the theory of probability. Among them was the idea that probability is an undefinable logical relationship between one set of propositions and another, and that this relation is associated with the *rational* degree of belief in a proposition *a* on the basis of others represented by *h*. Keynes had this to say:

> Let our premises consist of any set of propositions *h* and our conclusion consist of any set of propositions *a*, then if a knowledge of *h* justifies a rational degree of belief in *a* of degree *A*, we may say that there is a *probability-relation* of degree *A* between *a* and *h*.[27]

But Keynes also assumed (1) that not all degrees of belief *A* are numerically measurable, and (2) that not all degrees of belief are even comparable.[28] These assumptions should avoid many of the difficulties with which Venn struggled, as did other writers who had pointed to the difficulty of measuring degrees of belief numerically, and who claimed that the notion of degree of belief was psychological and not logical.

In the work of Ramsey (1926)[29] the notion of degree of belief is related to the notion of utility and to overt choices. He thought, as did Borel (1924)[30] that the only theoretically sound way of measuring a person's degree of belief is by examining his overt behavior—for example, the bet that he is prepared to make on the occurrence of an event. Most subjectivists have followed Ramsey and Borel in this, and identified degrees of belief with specific kinds of behavior such as making a certain wager, or choosing to stake a possible gain on the occurrence of one event rather than another.

(27) Keynes [1921], *A Treatise on Probability*, p. 4.
(28) *Ibid.*, p. 34.
(29) Ramsey, "Truth & Probability," reprinted here, and in Ramsey [1950], *The Foundations of Mathematics*.
(30) Borel [1924], [1964], "Apropos of a Treatise on Probability."

These are overt, behavioristic criteria: if a person would just as soon risk a gain on the toss of a coin that he is convinced is fair as on the occurrence of rain tomorrow, the degree of belief he has in the proposition that it will rain tomorrow is exactly one-half. Degree of belief, for most subjectivists, consists of a disposition to make certain specific kinds of choices in objectively defined choice situations. This is all very carefully discussed in Savage's *Foundations of Statistics.*[31]

Koopman (1940)[32] following Keynes, retains an intuitive notion of degree of belief. For both Koopman and Keynes, probabilities are not necessarily completely ordered, while *measurable* degrees of the same thing obviously must be. Some degrees of belief, according to these writers, could be assessed by the examination of betting behavior or choice under uncertainty, but other degrees of belief might turn out to be incomparable to these degrees in order of magnitude. Thus, in Koopman's view, it would make perfectly good sense to say that S is not more probable than T, and T is not more probable than S, without saying that they have the same probability. This seems to reflect a certain lack of order that we intuitively discover in our own opinions: it might be very hard for one to say, after having listened to the weather report and looked out the window, whether one has a smaller or a greater degree of belief in the proposition that it will rain tomorrow

(31) Savage [1954], *Foundations of Statistics*: The concepts of utility and probability have been related to one another in a number of subjectivist theories. Ramsey provided a complicated definition of a cardinal utility scale in terms of a special event (with probability 1/2) and then on the basis of this utility scale characterized the general notion of "degree of belief." See the article reprinted here, especially pages 25-52. His ideas are presented in a more formal manner in Davidson and Suppes [1956], "A Finite Axiomatization of Subjective Probability and Utility." Savage has shown how, on the basis of a qualitative ordering of preferences, a quantitative probability measure can be introduced. For a review of the history of the notion of riskless utility, see Stigler [1950], "The Development of Utility Theory." For reviews of the literature regarding the measurement of utility in situations involving risk, see Allais [1953], "Le comportement de l'homme rationel devant la risque: critique de l'ecole Americaine," Arrow [1958], "Utilities, Attitudes, Choices: A Review Article," and Friedman and Savage [1952], "The Utility Analysis of Choice Involving Risk." The classic paper in the field is D. Bernoulli [1731], "Specium Theoriae Novae De Mensura Sortis."

(32) In the article reprinted here and in the more technical articles, Koopman [1940b], "The Axioms and Algebra of Intuitive Probability," and Koopman [1941], "Intuitive Probabilities and Sequences."

than in the proposition that the coin about to be tossed will land heads; the degrees of the two beliefs are simply *different*.

A number of writers in recent years have suggested ways of characterizing beliefs which accommodate this problem. Kyburg[33] uses pairs of real numbers to characterize beliefs; Good[34] considers a series of probability judgments, in which one considers a first level probability judgment to be more or less probable, which is a second level probability judgment; and then may consider how probable that judgment is, and so on. He also employs an interval representation for each level of probability judgment. Isaac Levi[35] has developed a representation of belief as a convex set of probability functions; this is more general than an interval representation. Among statisticians, Dempster,[36] C. A. B. Smith,[37] and Shafer,[38] have developed ways of characterizing beliefs that admit more structure than the betting ratio characterization. These approaches all pay a certain cost in greater complexity than the views represented here, and not all fall squarely in the subjectivistic camp. They are mentioned to indicate that the notion of degree of belief is not a simple one, and that most writers on subjectivistic probability employ a frankly idealized notion.

The notion of *coherence* of a body of beliefs was first introduced explicitly by Ramsey in 1926. As we mentioned above, Ramsey took overt behavior in situations of choice as indicating degrees of belief; in particular he took the least odds at which a person was willing to bet on a proposition as indicating his degree of belief in that proposition. If these odds are *r:s,* then the probability of that proposition for him (his degree of belief in it) is $r/(r + s)$. In his discussion of the bets that could be placed on each of a set of related propositions, Ramsey mentioned one important constraint: *coherence.* No set of bets on a set of propositions is allowable which insures that whatever the outcome of events, the bettor will lose money. It is possible to show that conformity of degrees of belief to the rules of the probability calculus is a necessary and sufficient condition of coherence in this sense. The restriction to coherence thus formulates a natural criterion of rationality in situations of uncertainty. Rationality is used in a normative sense here; coherence formulates a criterion of how a person's degrees of belief *ought* to be related.

(33) Kyburg [1961], *Probability and the Logic of Rational Belief*; Kyburg [1974], *The Logical Foundations of Statistical Inference.*
(34) In the paper reprinted here, and elsewhere.
(35) Levi [1974], "On Indeterminate Probabilities."
(36) Dempster [1967b], "Upper and Lower Probabilities Induced by a Multivalued Mapping."
(37) Smith [1965], "Personal Probability and Statistical Analysis."
(38) Shafer [1976], *A Mathematical Theory of Evidence.*

The notion of *coherence* has undergone some development in the hands of subjectivists. Two senses are now current. The weaker sense, due independently to Ramsey and de Finetti and employed also by Lehman,[39] is that just described. It requires that there be no set of bets at odds determined by the person's degrees of belief under which he is *bound* to lose whatever the state of affairs of the world. "Coherence" is also used in a stronger sense—*strict coherence*—due to Abner Shimony[40] (and employed also by John Kemeny[41]) in which the demand is not only made that it be impossible for book to be made against the holder of a coherent set of beliefs, but also that it be impossible to choose bets and stakes in such a way that (a) there is no chance that the holder of the beliefs will win a net amount, and also (b) there is a chance that he will suffer a net loss. In the case of the weak definition of coherence, the bettor is incoherent if he so bets that he *must* lose money, while in the case of the strong definition of coherence, the bettor is incoherent if he bets in such a way that he *may* either come out even or lose money. For example, Tom's beliefs could be such that he would accept an even-money bet that a certain random quantity will turn out to be less than M, and such that he would accept an even-money bet that it will turn out to be more than M, but that he regards it as infinitely improbable that it will turn out to be exactly M. A shrewd bookie will make an even-money bet with him that $X \leqslant M$, and also an even-money bet with him that $X \geqslant M$. Now it is quite true that if $X < M$, or if $X > M$, he will lose one of these bets and Tom will lose one, so that they will break even. But if $X = M$, then Tom will lose both bets. This violates the condition of strict coherence. The requirement of strict coherence leads to a somewhat stronger set of axioms for probability than does the requirement of weak coherence used by de Finetti.

Some English and American writers use the term "consistency" for "coherence."[42] From a logical point of view there are good arguments against this usage. In judging beliefs, "consistency" has a definite traditional meaning: beliefs are consistent if they don't contradict each other. To introduce a new meaning for this term seems highly inadvisable, particularly when such a nice word as "coherence" is ready at hand, waiting to be given a technical meaning.

It is clear from the previous discussion that the notion of "coherence" is intimately tied up with the normative character of the subjectivistic theory of probability. There has been some attempt on the part of psychologists, mathematicians, and logicians to determine to what extent the actual

(39) Lehman [1955], "On Confirmation and Rational Betting."
(40) Shimony [1955], "Coherence and the Axioms of Confirmation."
(41) Kemeny [1955], "Fair Bets and Inductive Probabilities."
(42) For example, Savage, Hewitt, and Good.

behavior of people conforms to the normative requirements of subjectivistic probability. A number of interesting articles and books have been written on the subject. One of the leading workers in the field is W. Edwards,[43] but the works of Davidson, Suppes, and Siegel[44] as well as articles by de Finetti,[45] Fréchet,[46] Tversky,[47] and Holstein[48] may also be mentioned.

In some ways the most important concept of the subjectivistic theory is that of *exchangeable events.* Until this notion was introduced by de Finetti in 1931, the subjectivistic theory of probability remained pretty much of a philosophical curiosity. None of those for whom probability theory was a means of livelihood or knowledge paid much attention to it. But with the introduction of the concept of "equivalence," or "symmetry," or "exchangeability," as it is now known, a way was discovered to connect the notion of subjective probability with the classical procedures of statistical inference.

In general the connection takes this form: it is shown that regardless of the subjective probability assignments with which a person starts (or rather, within very wide limits of such assignments), the results of applying the procedures of Bayesian statistical inference are the same: the same action is to be taken, the same hypothesis accepted, the same (or nearly the same) value of a parameter adopted. Bayesian statistical inference here is just the classical form of statistical inference according to which one starts with known initial probability distributions (*a priori* probabilities) and modifies these in the light of experiment and experience; it is to be distinguished from the procedures of the English-American school, which establish rules for the acceptance and rejection of statistical hypotheses and courses of action simply in accordance with statistical tests that have certain attractive properties.

Exchangeability covers a wide variety of situations, and functions in just this way: in the case of exchangeable events certain types of inference are relatively independent of the original assignment of probabilities to the individual events of a sequence. In particular de Finetti shows that the

(43) Edwards [1954], "The Theory of Decision Making," and Edwards [1961], "Behavioral Decision Theory."
(44) Davidson, Suppes, and Siegel [1957], *Decision Making.*
(45) de Finetti [19611], "Dans quel sens la theorie des decisions est-elle et doit-elle etre normative?"
(46) Fréchet [1954], "Une probleme psychologique sur les probabilités subjectives irrationelles," and [1955], "Sur l'importnace en économétrie de le distinction entre les probabilités rationelles et irrationelles."
(47) Tversky [1975], "A Critique of Expected Utility Theory; Descriptive and Normative Considerations."
(48) Holstein [1973], "The Concept of Probability in Psychological Experiments."

classical limit theorems on which many forms of statistical inference depend hold just as well for sequences of exchangeable events as they do for the sequences of independent, equiprobable events to which they have traditionally been applied. He shows, for example, that in the case of a sequence of exchangeable events, a person, whatever be the opinion with which he starts out, must, if he is to be coherent in his beliefs, after a sufficient amount of observation come to assign a probability to the type of event in question which is close to its observed relative frequency.

What, precisely, is exchangeability? An example may make it clear. Consider such sequences of events as tosses of a coin, throws of a die, measurements of a bean drawn at random from a barrel of beans, etc. On an empirical view, such sequences are composed of *independent* events or measurements: independent because there is no empirical or causal connection between one measurement or event and another. Thus, drawing beans from an urn of unknown composition one by one, replacing each bean after it is drawn, the objectivist would say, "The probability that the second bean is black is independent of the fact that the first bean is black; there is no causal relationship between the color of the first bean and the color of the second." For him the conditional probability that the second bean is black, *given* that the first bean is black, is just the same as the absolute probability that the second bean is black. It is also the same as the probability that the first bean is black. But all these probabilities are unknown.

The subjectivist will deny this. Although the probability that the $(i + 1)$st bean is black might have the value p (say, $1/2$) for him, the *conditional* probability (the probability that the $(i + 1)$st bean is black, *given* that the first i beans have all turned out to be black) is very likely to have a different value. The events in the sequence of drawings are not at all independent for the subjectivist; they are not independent for knowledge or degrees of belief. The fact that certain events have occurred in the sequence gives us evidence about the occurrence of future events which affects our degrees of belief about them. The conditional probability that I attribute to the $(i + 1)$st bean's being black, given knowledge of the color of the first i beans, is obviously dependent on the proportion of black beans I have observed in the first i draws. But if the drawings are "fair" this probability is equally obviously not dependent on the particular *order* in which the black beans appeared on the first i drawings. This is the important part of the objectivist's assertion of independence, and it is the crux of the notion of exchangeability.

The matter can be put more formally. Consider a sequence of trials, to the outcome of each of which we assign a number X; we assign the number X_i to the outcome of the ith trial. (For example, consider a sequence of tosses of a coin; we let $X_i = 0$ if the ith toss results in heads, $X_i = 1$ otherwise.) Let us write $P(X_i)$ for the probability distribution of X_i, and

$P(X_i | X_{j_1} \ldots X_{j_n})$ for the conditional probability distribution of X_i, given that the j_1th trial resulted in the number X_{j_1}, that the j_2th trial resulted in the number X_{j_2}, etc. If we give P an objectivist interpretation, such events as tosses of a coin are *independent* (and equiprobable) in view of the fact that $P(X_i | X_{j_1} \ldots X_{j_n}) = P(X_i)$ for all distinct subscripts $(i, j_1, \ldots j_n)$. If we give P a subjectivist interpretation, these events are exchangeable (in the opinions of most people—it is a matter of subjective probability assignments), in view of the fact that although $P(X_i) \neq P(X_i | X_{j_1} \ldots X_{j_2})$ (or else we would be unable to learn from experience), $P(X_i) = P(X_j)$, and more important, $P(X_i | X_{j_1} \ldots X_{j_n}) = P(X_h | X_{k_1} \ldots X_{k_n})$ for any sets of distinct subscripts $(i, j_1, \ldots j_n)$ and $(h, k_1, \ldots k_n)$ provided that the same set of numbers (of 0's and 1's) occur to the right of the vertical bar in either case. In other words, the probability of heads on the tenth pass of a coin, given that the first two tosses resulted heads and the next seven resulted in tails, will be just the same as the probability of heads on the tenth toss of a coin given that the twenty-first through twenty-seventh result in tails, and that the fifteenth and sixteenth tosses result in heads. Exchangeable events are events that occur in a sequence in random manner; the order of their occurrence does not affect the probabilities we are interested in.

The notion of exchangeability was first introduced by de Finetti (1931). It forms an interesting chapter in probability theory even for those who do not share the subjectivist point of view, and many mathematicians have worked with it for its own sake. Among those who have worked on the technical mathematics of exchangeability are Khinchin,[49] de Finetti,[50] Hewitt and Savage,[51] Ryll-Nardzewski,[52] and Diaconos and Freedman.[53] The theorem, known in various forms as "de Finetti's representation theorem," on which these crucial arguments depend, is this: A set of distributions for exchangeable random quantities can be represented as a probabilistic mixture of distributions for the same random quantities construed as independent and identically distributed. This is the mathematical link between classical statistical theory and subjectivistic Bayesian theory.

(49) Khinchin [1932]; "Sur les classes d'evenements equivalents," and Khinchin [1952], "On Classes of Equivalent Events."

(50) de Finetti [1938], "Sur la condition d'equivalence partielle."

(51) Hewitt and Savage [1955], "Symmetric Measures on Cartesian Products."

(52) Ryll-Nardzewski [1957], "On Stationary Sequences of Random Variables and de Finetti's Equivalences."

(53) Diaconos [1977] "Finite Forms of de Finetti's Theorem of Exchangeability," and Diaconos and Freedman [1978], "De Finetti's Generalization of Exchangeability."

The three concepts discussed above are essential to the subjectivistic point of view concerning probability. Other terms may be introduced for particular purposes, but these three represent the most important concepts in the historical development of the theory, and they are also the most essential concepts for understanding the general subjectivistic arguments that follow.

There have been a number of developments in this area in the past several years, and a number of issues have become the subject of controversy and debate.

One line of development has been the attempt to recognize the validity of Keynes' and Koopman's idea that not all probabilities are comparable. C. A. B. Smith, in a number of publications,[54] has sought to give this idea expression in a subjectivistic framework by taking the fundamental behavioral notion as the willingness to offer odds on a proposition. Where Ramsey, Savage, and others stipulated that one's degrees of belief in a proposition should be represented by the odds at which one would take *either side* of a bet on that proposition, Smith points out that the odds one offers in favor of a proposition and the odds one offers against a proposition need not be the inverse of one another: if I am willing to offer 3:1 odds for P (corresponding to a betting ratio of 3/4) I will not generally offer odds of 1:3 against P (corresponding to a betting ratio of 1/4); rather, I may offer odds of 1:4 against P (corresponding to a betting ratio of 1/5). We thus arrive at upper and lower probabilities for a proposition, determined respectively by the least odds I will offer against the proposition and the least odds I will offer in favor of it. The idea of upper and lower probabilities, first developed for a logical interpretation of probability by Kyburg, has been independently developed in one form or another by various writers on subjective probability in addition to Smith: Good,[55] Shafer,[56] Dempster,[57] and others.

A further development of this idea is due to Levi.[58] Given an algebra of propositions, we may consider not only the upper and lower probabilities of each of these propositions, but a whole set of subjective probability functions defined over the algebra. We stipulate that this set of probability functions be convex: that is, that the weighted average of any two of them also belongs to the set. This representation of belief has the consequence that there are upper and lower probabilities associated with each proposition; but it also provides a richer structure for the representation of conditional beliefs as concretized in conditional bets.

(54) Smith [1965], "Personal Probability and Statistical Analysis."
(55) I. J. Good, this volume.
(56) See footnote 38, this article.
(57) See footnote 36, this article.
(58) Levi [1974], "On Indeterminate Probabilities."

A number of statisticians have become advocates of the subjectivistic point of view. The best known and most ardent is perhaps D. V. Lindley, whose two volume text[59] is among the most influential modern books in the area. The point of view has also been adopted by decision theorists (Schlaifer, Raiffa, Luce)[60] and through them been introduced into economics, political science, and business management. At the same time, psychologists (e.g., Edwards[61]) have attempted to improve upon the theory—that is, to make it more realistic. It was pointed out earlier that as ordinarily construed, the theory does not accurately reflect the behavior of ordinary people—particularly with regard to expectations based on very rare events. Efforts have been made to modify the theory to reflect more accurately the actual choice behavior of individuals in situations of risk.

Three main technical issues have come to the fore in the discussion of subjective probability. The first is the distinction between strict coherence and ordinary coherence. This is connected with the concept of regularity, so-called because Carnap proposed an axiom of regularity for his system of logical probability. The axiom of regularity stipulates that only statements that are logically certain should bear the probability value 1, and that only statements that are logically impossible should bear the probability value 0. This is correlated with the idea of strict coherence. Regularity implies strict coherence. A system of beliefs is strictly coherent only if every proposition assigned a probability of 0 is logically impossible: there should be no set of bets reflecting one's beliefs under which it is certain that one cannot win, and possible that one can lose. Another issue which arises concerns countable additivity. A standard axiom for subjective probability requires that if P is equivalent to a disjunction of a finite number of logically exclusive alternatives, the probability of P is to be the sum of the probabilities of those alternatives. The question is whether or not to extend this axiom to apply also to a countable number of alternatives. De Finetti[62] and others have argued strongly against the adoption of countable additivity; but countable additivity is very important in classical statistics and probability theory.

There is no theorem connecting countable additivity and regularity. It is possible that both can be satisfied: A distribution of probabilities over the positive integers given by the function $P(i) = 1/2^i$ requires countable additiv-

(59) Lindley [1965a,b], *Introduction to Probability and Statistics. Part 1. Probability,* and *Part 2. Inference.*

(60) Raiffa and Schlaifer [1961], *Applied Statistical Decision Theory,* and Luce and Raiffa [1957], *Games and Decisions: Introduction and Critical Survey.*

(61) Edwards [1960], "Measurement of Utility and Subjective Probability."

(62) de Finetti [1972], *Probability, Induction, and Statistics.*

ity, but also satisfies regularity. Many writers (implicitly) reject regularity but accept countable additivity, by supposing that "data" is given probability 1, and by accepting countable additivity as an axiom. It would be perfectly possible to accept regularity (refusing to assign probability 1 to any statement not logically certain), and to reject countable additivity. And of course one can reject both, as one might well if one were inclined to suppose that there is a probability function over the integers that took them to be "equally probable."

The third issue, involved more with statistical applications to the theory than with the philosophical issues, turns on the use of so-called improper priors in statistical inference. Although the notion of exchangeability holds for most kinds of statistical evidence, and thus generally guarantees that the particular prior probability distribution one starts with is of relatively little importance in statistical inference, nevertheless one does have to start with *some* prior, and the particular one one starts with does make some diffence. An improper prior distribution is one whose integral over the whole space of possibilities is infinite in value, instead of having the value 1. Plausible results in statistical inference can be obtained through the use of improper priors to express such judgments as: the ratio of white balls to black balls in this urn may have any value from 0 to 1, and so far as I am concerned, one value is no more probable than another; or, the mean of this normal distribution may have any value whatever, and I regard all values as equally probable. A single improper prior causes no formal difficulty; the posterior distribution, conditional on some evidence, is perfectly well behaved; the statistician is off and running. But sometimes the use of improper priors for different quantities can lead to anomalies and even, in some instances, to a violation of the principle of conditionalization which lies at the heart of subjectivistic statistics; it may turn out that your degree of belief depends on the order in which you consider two parts of your evidence.

One response to such anomalies on the part of some Bayesians has been to attempt to offer arguments for using prior distributions satisfying certain formal constraints. These arguments (for example, those of Rosenkrantz and Jaynes[63]) have not been generally regarded as convincing, and indeed there are serious difficulties with them, as Seidenfeld[64] has pointed out.

These two important issues—the issue of regularity, or strict coherence, and the issue of improper priors—have generated a considerable literature, both in statistics and in philosophical probability theory. They are both

(63) Jaynes [1968], "Prior Probabilities." Rosenkrantz [1977], *Inference, Method, & Decision.*
(64) Seidenfeld [1979], "Why I Am Not an Objective Bayesian."

relatively technical issues, however, and though there are many papers pertaining to them that could be collected and would make interesting reading, we decided against including any of them here.

The subjectivist theory has undergone a fair amount of criticism, especially as it has entered the mainstream of thought concerning probability in the last twenty years. There are those who argue that inductive reasoning must involve more norms of rationality than are built into the subjectivist conception of probability. However internally consistent the subjectivist theory is, it fails to meet the standards of rationality which we demand; it neither guarantees uniformity of reasoning between persons at a given time or even that shifts in belief over time have similar effects for different persons. This seems the basis for Carnap's disagreement with the subjectivists with whom he is in agreement on so many other issues. Both Kyburg[65] and Levi[66] feel that room must be left for a principle of direct inference.

A criticism closer to home is that no really consistent account can be given of the theory. Our discussion has already made clear that the subjectivist theory of probability relates probabilities (degrees of belief of persons) to preferences (choices of persons), i.e., that it can be construed as a theory of degrees of belief or as a theory of decisions. Furthermore, each of these theories can be interpreted as a normative or as an empirical theory, and so we get a fourfold interpretation of the subjectivist theory:

	empirical	normative
Decision theory	empirical decison theory	normative decision theory
Theory of belief	empirical theory of belief	normative theory belief

Much evidence has revealed the falsity of the theory if it is taken to be either an empirical decision theory or an empirical theory of belief.[67] But in order to treat it as a normative theory, very unrealistic assumptions must be introduced as to the conditions under which a person should accept bets (in the case of normative decision theory) or as to the *a priori* character of certain critical assignments of degrees of belief (in the case of a normative theory of belief). All these arguments are spelled out in Kyburg's recent article on

(65) Kyburg [1977], "Randomness and the Right Reference Class."
(66) Levi [1977], "Direct Inference."
(67) Edwards [1960], "Measurement of Utility and Subjective Probability."
        Slavic et al. [1977], "Behavioral Decision Theory."

the subject.[68] The theory of subjective probability has undergone extensive developments in recent years, and is both more widely known and more widely held than it was ten or fifteen years ago. Despite the hopes of its most ardent adherents, it has not swept the field of statistics—nor of business management, philosophy, psychology, or social science—by storm. Few people, in fact, are whole-heartedly commited to the theory, although there are a significant number who find some place for the theory in their arsenal of weapons for attacking problems of statistical inference, epistemology, or decision theory. It has become a respectable minority view. It is hard to forecast what the future will bring, but it is certainly safe to say that the theory has achieved a firm historical niche, and is worthy of serious study.

(68) Kyburg [1978a], "Subjective Probability: Criticisms, Reflections and Problems."

FRANK PLUMPTON RAMSEY

Truth and Probability

(1926)

This selection is reprinted from the author's *The Foundations of Mathematics and other Logical Essays* (Ed. R. B. Braithwaite), The Humanities Press, New York, 1950. Reprinted by permission of The Humanities Press and of Routledge & Kegan Paul Ltd.

FOREWORD

In this essay the Theory of Probability is taken as a branch of logic, the logic of partial belief and inconclusive argument; but there is no intention of implying that this is the only or even the most important aspect of the subject. Probability is of fundamental importance not only in logic but also in statistical and physical science, and we cannot be sure beforehand that the most useful interpretation of it in logic will be appropriate in physics also. Indeed the general difference of opinion between statisticians who for the most part adopt the frequency theory of probability and logicians who mostly reject it renders it likely that the two schools are really discussing different things, and that the word "probability" is used by logicians in one sense and by statisticians in another. The conclusions we shall come to as to the meaning of probability in logic must not, therefore, be taken as prejudging its meaning in physics.[1]

1. The Frequency Theory. In the hope of avoiding some purely verbal controversies, I propose to begin by making some admissions in favour of the frequency theory. In the first place this theory must be conceded to have a firm basis in ordinary language, which often uses "probability" practically as a synonym for proportion; for example, if we say that the probability of recovery from smallpox is three-quarters, we mean, I think, simply that that is the proportion of smallpox cases which recover. Secondly, if we start with what is called the calculus of probabilities, regarding it first as a branch of pure mathematics, and then looking round for some interpretation of the formulae which shall show that our axioms are consistent and our subject not entirely useless, then much the simplest and least controversial interpretation of the calculus is one in terms of frequencies. This is true not only of the ordinary mathematics of probability, but also of the symbolic calculus developed by Mr. Keynes; for if in his a/h, a and h are taken to be not propositions but propositional functions or class-concepts which define finite classes, and a/h is taken to

---

[1] [A final chapter, on probability in science, was designed but not written.–R.B.]

25

mean the proportion of members of *h* which are also members of *a,* then all his propositions become arithmetical truisms.

Besides these two inevitable admissions, there is a third and more important one, which I am prepared to make temporarily although it does not express my real opinion. It is this. Suppose we start with the mathematical calculus, and ask, not as before what interpretation of it is most convenient to the pure mathematician, but what interpretation gives results of greatest value to science in general, then it may be that the answer is again an interpretation in terms of frequency; that probability as it is used in statistical theories, especially in statistical mechanics—the kind of probability whose logarithm is the entropy—is really a ratio between the numbers of two classes, or the limit of such a ratio. I do not myself believe this, but I am willing for the present to concede to the frequency theory that probability as used in modern science is really the same as frequency.

But, supposing all this admitted, it still remains the case we have the authority both of ordinary language and of many great thinkers for discussing under the heading of probability what appears to be quite a different subject, the logic of partial belief. It may be that, as some supporters of the frequency theory have maintained, the logic of partial belief will be found in the end to be merely the study of frequencies, either because partial belief is definable as, or by reference to, some sort of frequency, or because it can only be the subject of logical treatment when it is grounded on experienced frequencies. Whether these contentions are valid can, however, only be decided as a result of our investigation into partial belief, so that I propose to ignore the frequency theory for the present and begin an inquiry into the logic of partial belief. In this, I think, it will be most convenient if, instead of straight away developing my own theory, I begin by examining the views of Mr. Keynes, which are so well known and in essentials so widely accepted that readers probably feel that there is no ground for re-opening the subject *de novo* until they have been disposed of.

2. Mr. Keynes' Theory. Mr. Keynes[1] starts from the supposition that we make probable inferences for which we claim objective validity; we proceed from full belief in one proposition to partial belief in another, and we claim that this procedure is objectively right, so that if another man in similar circumstances entertained a different degree of belief, he would be wrong in doing so. Mr. Keynes accounts for this by supposing that between any two propositions, taken as premiss and conclusion, there holds one and only one relation of a certain sort called probability relations; and that if, in any given case, the relation is that of degree *a,* from full belief in the

---

(1) J. M. Keynes, *A Treatise on Probability,* 1921.

premiss, we should, if we were rational, proceed to a belief of degree *a* in the conclusion.

Before criticizing this view, I may perhaps be allowed to point out an obvious and easily corrected defect in the statement of it. When it is said that the degree of the probability relation is the same as the degree of belief which it justifies, it seems to be presupposed that both probability relations, on the one hand, and degrees of belief on the other can be naturally expressed in terms of numbers, and then that the number expressing or measuring the probability relation is the same as that expressing the appropriate degree of belief. But if, as Mr. Keynes holds, these things are not always expressible by numbers, then we cannot give his statement that the degree of the one is the same as the degree of the other such a simple interpretation, but must suppose him to mean only that there is a one-one correspondence between probability relations and the degrees of belief which they justify. This correspondence must clearly preserve the relations of greater and less, and so make the manifold of probability relations and that of degrees of belief similar in Mr. Russell's sense. I think it is a pity that Mr. Keynes did not see this clearly, because the exactitude of this correspondence would have provided quite as worthy material for his scepticism as the numerical measurement of probability relations. Indeed some of his arguments against their numerical measurement appear to apply quite equally well against their exact correspondence with degrees of belief; for instance, he argues that if rates of insurance correspond to subjective, i.e. actual, degrees of belief, these are not rationally determined, and we cannot infer that probability relations can be similarly measured. It might be argued that the true conclusion in such a case was not that, as Mr. Keynes thinks, to the non-numerical probability relation corresponds a non-numerical degree of rational belief, but that degrees of belief, which were always numerical, did not correspond one to one with the probability relations justifying them. For it is, I suppose, conceivable that degrees of belief could be measured by a psychogalvanometer or some such instrument, and Mr. Keynes would hardly wish it to follow that probability relations could all be derivatively measured with the measures of the beliefs which they justify.

But let us now return to a more fundamental criticism of Mr. Keynes' views, which is the obvious one that there really do not seem to be any such things as the probability relations he describes. He supposes that, at any rate in certain cases, they can be perceived; but speaking for myself I feel confident that this is not true. I do not perceive them, and if I am to be persuaded that they exist it must be by argument; moreover I shrewdly suspect that others do not perceive them either, because they are able to come to so very little agreement as to which of them relates any two given propositions. All we appear to know about them are certain general propositions, the laws of addition and multiplication; it is as if everyone knew the laws of geometry but no one could tell whether any given object were round or square; and I find it hard to imagine how so large a body of general knowledge can be combined with so slender a stock of particular facts. It is true that about some particular cases there is agreement, but these somehow paradoxically are always immensely complicated; we all agree that the probability of a coin coming down heads is $\frac{1}{2}$, but we can none of us say exactly what is the evidence which forms the other term for the probability relation about which we are then judging. If, on the other hand, we take the simplest possible pairs of propositions such as "This is red" and "That is blue" or "This is red" and "That is red," whose logical relations should surely be easiest to see, no one, I think, pretends to be sure what is the probability relation which connects them. Or, perhaps, they may claim to see the relation but they will not be able to say anything about it with certainty, to state if it is more or less than $\frac{1}{3}$, or so on. They may, of course, say that it is incomparable with any numerical relation, but a relation about which so little can be truly said will be of little scientific use and it will be hard to convince a sceptic of its existence. Besides this view is really rather paradoxical; for any believer in induction must admit that between "This is red" as conclusion and "This is round," together with a billion propositions of the form "a is round and red" as evidence, there is a finite probability relation; and it is hard to suppose that as we accumulate instances there is suddenly a point, say after 233 instances, at which the probability relation becomes finite and so comparable with some numerical relations.

It seems to me that if we take the two propositions "a is red," "b is red," we cannot really discern more than four simple logical relations between them; namely identity of form, identity of predicate, diversity of subject, and logical independence of import. If anyone were to ask me what probability one gave to the other, I should not try to answer by contemplating the propositions and trying to discern a logical relation between them, I should, rather, try to imagine that one of them was all that I knew, and to guess what degree of confidence I should then have in the other. If I were able to do this, I might no doubt still not be content with it but might say, "This is what I should think, but, of course, I am only a fool" and proceed to consider what a wise man would think and call that the degree of probability. This kind of self-criticism I shall discuss later when developing my own theory; all that I want to remark here is that no one estimating a degree of probability simply contemplates the two propositions supposed to be related by it; he always considers *inter alia* his own actual or hypothetical degree of belief. This remark seems to me to be borne out by observation of my own behaviour; and to be the only way of accounting for the fact that we can all give estimates of probability in cases taken from actual life, but are quite unable to do so in the logically simplest cases in which, were probability a logical relation, it would be easiest to discern.

Another argument against Mr. Keynes' theory can, I think, be drawn from his inability to adhere to it consistently even in discussing first principles. There is a passage in his chapter on the measurement of probabilities which reads as follows:

> Probability is, *vide* Chapter II (§ 12), relative in a sense to the principles of *human* reason. The degree of probability, which it is rational for *us* to entertain, does not presume perfect logical insight, and is relative in part to the secondary propositions which we in fact know; and it is not dependent upon whether more perfect logical insight is or is not conceivable. It is the degree of probability to which those logical processes lead, of which our minds are capable; or, in the language of Chapter II, which those secondary propositions justify, which we in fact know. If we do not take this view of probability, if we do not limit it in this way and make it, to this extent, relative to human powers, we are altogether adrift in the unknown; for we cannot ever know what degree of probability would be justified by the perception of logical relations which we are, and must always be, incapable of comprehending.[2]

This passage seems to me quite unreconcilable with the view which Mr. Keynes adopts everywhere except in this and another similar passage. For he generally holds that the degree of belief which we are justified in placing in the conclusion of an argument is determined by what relation of probability unites that conclusion to our premisses. There is only one such relation and consequently only one relevant true secondary proposition, which, of course, we may or may not know, but which is necessarily independent of the human mind. If we do not know it, we do not know it and cannot tell how far we ought to believe the conclusion. But often, he supposes, we do know it; probability relations are not ones which we are incapable of comprehending. But on this view of the matter the passage quoted above has no meaning: the relations which justify probable beliefs are probability relations, and it is nonsense to speak of them being justified by logical relations which we are, and must always be, incapable of comprehending.

The significance of the passage for our present purpose lies in the fact that it seems to presuppose a different view of probability, in which indefinable probability relations play no part, but in which the degree of rational belief depends on a variety of logical relations. For instance, there might be

(2) *Ibid.*, p. 32, his italics.

between the premiss and conclusion the relation that the premiss was the logical product of a thousand instances of a generalization of which the conclusion was one other instance, and this relation, which is not an indefinable probability relation but definable in terms of ordinary logic and so easily recognizable, might justify a certain degree of belief in the conclusion on the part of one who believed the premiss. We should thus have a variety of ordinary logical relations justifying the same or different degrees of belief. To say that the probability of a given h was such-and-such would mean that between a and h was some relation justifying such-and-such a degree of belief. And on this view it would be a real point that the relation in question must not be one which the human mind is incapable of comprehending.

This second view of probability as depending on logical relations but not itself a new logical relation seems to me more plausible than Mr. Keynes' usual theory; but this does not mean that I feel at all inclined to agree with it. It requires the somewhat obscure idea of a logical relation justifying a degree of belief, which I should not like to accept as indefinable because it does not seem to be at all a clear or simple notion. Also it is hard to say what logical relations justify what degrees of belief, and why; any decision as to this would be arbitrary, and would lead to a logic of probability consisting of a host of so-called "necessary" facts, like formal logic on Mr. Chadwick's view of logical constants.[3] Whereas I think it far better to seek an explanation of this "necessity" after the model of the work of Mr. Wittgenstein, which enables us to see clearly in what precise sense and why logical propositions are necessary, and in a general way why the system of formal logic consists of the propositions it does consist of, and what is their common characteristic. Just as natural science tries to explain and account for the facts of nature, so philosophy should try, in a sense, to explain and account for the facts of logic; a task ignored by the philosophy which dismisses these facts as being unaccountably and in an indefinable sense "necessary."

Here I propose to conclude this criticism of Mr. Keynes' theory, not because there are not other respects in which it seems open to objection, but because I hope that what I have already said is enough to show that it is not so completely satisfactory as to render futile any attempt to treat the subject from a rather different point of view.

3. Degrees of Belief. The subject of our inquiry is the logic of partial belief, and I do not think we can carry it far unless we have at least an approximate notion of what partial belief is, and how, if at all, it can be measured. It will not be very enlightening to be told that in such circumstances it would be rational to believe a proposition to the extent of $\frac{2}{3}$, unless

(3) J. A. Chadwick, "Logical Constants," *Mind* (1927).

we know what sort of a belief in it that means. We must therefore try to develop a purely psychological method of measuring belief. It is not enough to measure probability; in order to apportion correctly our belief to the probability we must also be able to measure our belief.

It is a common view that belief and other psychological variables are not measurable, and if this is true our inquiry will be vain; and so will the whole theory of probability conceived as a logic of partial belief; for if the phrase "a belief two-thirds of certainty" is meaningless, a calculus whose sole object is to enjoin such beliefs will be meaningless also. Therefore unless we are prepared to give up the whole thing as a bad job we are bound to hold that beliefs can to some extent be measured. If we were to follow the analogy of Mr. Keynes' treatment of probabilities we should say that some beliefs were measurable and some not; but this does not seem to me likely to be a correct account of the matter: I do not see how we can sharply divide beliefs into those which have a position in the numerical scale and those which have not. But I think beliefs do differ in measurability in the following two ways. First, some beliefs can be measured more accurately than others; and, secondly, the measurement of beliefs is almost certainly an ambiguous process leading to a variable answer depending on how exactly the measurement is conducted. The degree of a belief is in this respect like the time interval between two events; before Einstein it was supposed that all the ordinary ways of measuring a time interval would lead to the same result if properly performed. Einstein showed that this was not the case; and time interval can no longer be regarded as an exact notion, but must be discarded in all precise investigations. Nevertheless, time interval and the Newtonian system are sufficiently accurate for many purposes and easier to apply.

I shall try to argue later that the degree of a belief is just like a time interval; it has no precise meaning unless we specify more exactly how it is to be measured. But for many purposes we can assume that the alternative ways of measuring it lead to the same result, although this is only approximately true. The resulting discrepancies are more glaring in connection with some beliefs than with others, and these therefore appear less measurable. Both these types of deficiency in measurability, due respectively to the difficulty in getting an exact enough measurement and to an important ambiguity in the definition of the measurement process, occur also in physics and so are not difficulties peculiar to our problem; what is peculiar is that it is difficult to form any idea of how the measurement is to be conducted, how a unit is to be obtained, and so on.

Let us then consider what is implied in the measurement of beliefs. A satisfactory system must in the first place assign to any belief a magnitude or degree having a definite position in an order of magnitudes; beliefs which are of the same degree as the same belief must be of the same degree as one another and so on. Of course this cannot be accomplished without introducing a certain amount of hypothesis or fiction. Even in physics we cannot maintain that things that are equal to the same thing are equal to one another unless we take "equal" not as meaning "sensibly equal" but a fictitious or hypothetical relation. I do not want to discuss the metaphysics or epistemology of this process, but merely to remark that if it is allowable in physics it is allowable in psychology also. The logical simplicity characteristic of the relations dealt with in a science is never attained by nature alone without any admixture of fiction.

But to construct such an ordered series of degrees is not the whole of our task; we have also to assign numbers to these degrees in some intelligible manner. We can of course easily explain that we denote full belief by 1, full belief in the contradictory by 0, and equal beliefs in the proposition and its contradictory by $\frac{1}{2}$. But it is not easy to say what is meant by a belief $\frac{2}{3}$ of certainty, or a belief in the proposition being twice as strong as that in its contradictory. This is the harder part of the task, but it is absolutely necessary; for we do calculate numerical probabilities, and if they are to correspond to degrees of belief we must discover some definite way of attaching numbers to degrees of belief. In physics we often attach numbers by discovering a physical process of addition:[4] the measure-numbers of lengths are not assigned arbitrarily subject only to the proviso that the greater length shall have the greater measure; we determine them further by deciding on a physical meaning for addition; the length got by putting together two given lengths must have for its measure the sum of their measures. A system of measurement in which there is nothing corresponding to this is immediately recognized as arbitrary, for instance Mohs' scale of hardness[5] in which 10 is arbitrarily assigned to diamond, the hardest known material, 9 to the next hardest, and so on. We have therefore to find a process of addition for degrees of belief, or some substitute for this which will be equally adequate to determine a numerical scale.

Such is our problem; how are we to solve it? There are, I think, two ways in which we can begin. We can, in the first place, suppose that the degree of a belief is something perceptible by its owner; for instance that beliefs differ in the intensity of a feeling by which they are accompanied, which might be called a belief-feeling or feeling of conviction, and that by the degree of belief we mean the intensity of this feeling. This view would be very inconvenient, for it is not easy to ascribe numbers to the intensities of feelings; but apart from this it seems to me observably false, for the beliefs

(4) See N. Campbell, *Physics, The Elements*, 1920, p. 277.
(5) *Ibid.*, p. 271.

which we hold most strongly are often accompanied by practically no feeling at all; no one feels strongly about things he takes for granted.

We are driven therefore to the second supposition that the degree of a belief is a causal property of it, which we can express vaguely as the extent to which we are prepared to act on it. This is a generalization of the well-known view, that the differentia of belief lies in its causal efficacy, which is discussed by Mr. Russell in his *Analysis of Mind*. He there dismisses it for two reasons, one of which seems entirely to miss the point. He argues that in the course of trains of thought we believe many things which do not lead to action. This objection is however beside the mark, because it is not asserted that a belief is an idea which does actually lead to action, but one which would lead to action in suitable circumstances; just as a lump of arsenic is called poisonous not because it actually has killed or will kill anyone, but because it would kill anyone if he ate it. Mr. Russell's second argument is, however, more formidable. He points out that it is not possible to suppose that beliefs differ from other ideas only in their effects, for if they were otherwise identical their effects would be identical also. This is perfectly true, but it may still remain the case that the nature of the difference between the causes is entirely unknown or very vaguely known, and that what we want to talk about is the difference between the effects, which is readily observable and important.

As soon as we regard belief quantitatively, this seems to me the only view we can take of it. It could well be held that the difference between believing and not believing lies in the presence or absence of introspectible feelings. But when we seek to know what is the difference between believing more firmly and believing less firmly, we can no longer regard it as consisting in having more or less of certain observable feelings; at least I personally cannot recognize any such feelings. The difference seems to me to lie in how far we should act on these beliefs: this may depend on the degree of some feeling or feelings, but I do not know exactly what feelings and I do not see that it is indispensable that we should know. Just the same thing is found in physics; men found that a wire connecting plates of zinc and copper standing in acid deflected a magnetic needle in its neighbourhood. Accordingly as the needle was more or less deflected the wire was said to carry a larger or a smaller current. The nature of this "current" could only be conjectured: what were observed and measured were simply its effects.

It will no doubt be objected that we know how strongly we believe things, and that we can only know this if we can measure our belief by introspection. This does not seem to me necessarily true; in many cases, I think, our judgment about the strength of our belief is really about how we should act in hypothetical circumstances. It will be answered that we can only tell how we should act by observing the present belief-feeling which determines how we should act; but again I doubt the cogency of the argument. It is possible that what determines how we should act determines us also directly or indirectly to have a correct opinion as to how we should act, without its ever coming into consciousness.

Suppose, however, I am wrong about this and that we can decide by introspection the nature of belief, and measure its degree; still, I shall argue, the kind of measurement of belief with which probability is concerned is not this kind but is a measurement of belief *qua* basis of action. This can I think be shown in two ways. First, by considering the scale of probabilities between 0 and 1, and the sort of way we use it, we shall find that it is very appropriate to the measurement of belief as a basis of action, but in no way related to the measurement of an introspected feeling. For the units in terms of which such feelings or sensations are measured are always, I think, differences which are just perceptible: there is no other way of obtaining units. But I see no ground for supposing that the interval between a belief of degree $\frac{1}{3}$ and one of degree $\frac{1}{2}$ consists of as many just perceptible changes as does that between one of $\frac{2}{3}$ and one of $\frac{5}{6}$, or a scale based on just perceptible differences would have any simple relation to the theory of probability. On the other hand the probability of $\frac{1}{3}$ is clearly related to the kind of belief which would lead to a bet of 2 to 1, and it will be shown below how to generalize this relation so as to apply to action in general. Secondly, the quantitative aspects of beliefs as the basis of action are evidently more important than the intensities of belief-feelings. The latter are no doubt interesting, but may be very variable from individual to individual, and their practical interest is entirely due to their position as the hypothetical causes of beliefs *qua* bases of action.

It is possible that some one will say that the extent to which we should act on a belief in suitable circumstances is a hypothetical thing, and therefore not capable of measurement. But to say this is merely to reveal ignorance of the physical sciences which constantly deal with and measure hypothetical quantities; for instance, the electric intensity at a given point is the force which would act on a unit charge if it were placed at the point.

Let us now try to find a method of measuring beliefs as bases of possible actions. It is clear that we are concerned with dispositional rather than with actualized beliefs; that is to to say, not with beliefs at the moment when we are thinking of them, but with beliefs like my belief that the earth is round, which I rarely think of, but which would guide my action in any case to which it was relevant.

The old-established way of measuring a person's belief is to propose a bet, and see what are the lowest odds which he will accept. This method I regard as fundamentally sound; but it suffers from being insufficiently general, and from being necessarily inexact. It is inexact partly because of the diminishing marginal utility of money, partly because the person may have a special eagerness or reluctance to bet, because he either enjoys or dislikes excitement or for any other reason, e.g. to make a book. The difficulty is like that of separating two different co-operating forces. Besides, the proposal of a bet may inevitably alter his state of opinion; just as we could not always measure electric intensity by actually introducing a charge and seeing what force it was subject to, because the introduction of the charge would change the distribution to be measured.

In order therefore to construct a theory of quantities of belief which shall be both general and more exact, I propose to take as a basis a general psychological theory, which is now universally discarded, but nevertheless comes, I think, fairly close to the truth in the sort of cases with which we are most concerned. I mean the theory that we act in the way we think most likely to realize the objects of our desires, so that a person's actions are completely determined by his desires and opinions. This theory cannot be made adequate to all the facts, but it seems to me a useful approximation to the truth particularly in the case of our self-conscious or professional life, and it is presupposed in a great deal of our thought. It is a simple theory and one which many psychologists would obviously like to preserve by introducing unconscious desires and unconscious opinions in order to bring it more into harmony with the facts. How far such fictions can achieve the required result I do not attempt to judge: I only claim for what follows approximate truth, or truth in relation to this artificial system of psychology, which like Newtonian mechanics can, I think, still be profitably used even though it is known to be false.

It must be observed that this theory is not to be identified with the psychology of the Utilitarians, in which pleasure had a dominating position. The theory I propose to adopt is that we seek things which we want, which may be our own or other people's pleasure, or anything else whatever, and our actions are such as we think most likely to realize these goods. But this is not a precise statement, for a precise statement of the theory can only be made after we have introduced the notion of quantity of belief.

Let us call the things a person ultimately desires "goods," and let us at first assume that they are numerically measurable and additive. That is to say that if he prefers for its own sake an hour's swimming to an hour's reading, he will prefer two hours' swimming to one hour's swimming and one hour's reading. This is of course absurd in the given case but this may only be because swimming and reading are not ultimate goods, and because we cannot imagine a second hour's swimming precisely similar to the first, owing to fatigue, etc.

Let us begin by supposing that our subject has no doubts about anything, but certain opinions about all propositions. Then we can say that he will always choose the course of action which will lead in his opinion to the greatest sum of good.

It should be emphasized that in this essay good and bad are never to be understood in any ethical sense but simply as denoting that to which a given person feels desire and aversion.

The question then arises how we are to modify this simple system to take account of varying degrees of certainty in his beliefs. I suggest that we introduce as a law of psychology that his behaviour is governed by what is called the mathematical expectation; that is to say that, if p is a proposition about which he is doubtful, any goods or bads for whose realization p is in his view a necessary and sufficient condition enter into his calculations multiplied by the same fraction, which is called the "degree of his belief in p." We thus define degree of belief in a way which presupposes the use of the mathematical expectation.

We can put this in a different way. Suppose his degree of belief in p is m/n; then his action is such as he would choose it to be if he had to repeat it exactly n times, in m of which p was true, and in the others false. [Here it may be necessary to suppose that in each of the n times he had no memory of the previous ones.]

This can also be taken as a definition of the degree of belief, and can easily be seen to be equivalent to the previous definition. Let us give an instance of the sort of case which might occur. I am at a cross-roads and do not know the way; but I rather think one of the two ways is right. I propose therefore to go that way but keep my eyes open for someone to ask; if now I see someone half a mile away over the fields, whether I turn aside to ask him will depend on the relative inconvenience of going out of my way to cross the fields or of continuing on the wrong road if it is the wrong road. But it will also depend on how confident I am that I am right; and clearly the more confident I am of this the less distance I should be willing to go from the road to check my opinion. I propose therefore to use the distance I would be prepared to go to ask, as a measure of the confidence of my opinion; and what I have said above explains how this is to be done. We can set it out as follows: suppose the disadvantage of going x yards to ask is $f(x)$, the advantage of arriving at the right destination is r, that of arriving at the wrong one w. Then if I should just be willing to go a distance d to ask, the degree of my belief that I am on the right road is given by

$$p = 1 - \frac{f(d)}{r - w}.$$

For such an action is one it would just pay me to take, if I had to act in the same way n times, in np of which I was on the right way but in the others not.

For the total good resulting from not asking each time

$$= npr + n(1 - p)w$$

$$= nw + np(r - w),$$

that resulting from asking at distance x each time

$$= nr - nf(x). \qquad \text{[I now always go right.]}$$

This is greater than the preceding expression, provided

$$f(x) < (r - w)(1 - p),$$

∴ the critical distance d is connected with p, the degree of belief, by the relation $f(d) = (r - w)(1 - p)$

$$\text{or } p = 1 - \frac{f(d)}{r - w} \qquad \text{as asserted above.}$$

It is easy to see that this way of measuring beliefs gives results agreeing with ordinary ideas; at any rate to the extent that full belief is denoted by 1, full belief in the contradictory by 0, and equal belief in the two by $\frac{1}{2}$. Further, it allows validity to betting as means of measuring beliefs. By proposing a bet on p we give the subject a possible course of action from which so much extra good will result to him if p is true and so much extra bad if p is false. Supposing the bet to be in goods and bads instead of in money, he will take a bet at any better odds than those corresponding to his state of belief; in fact his state of belief is measured by the odds he will just take; but this is vitiated, as already explained, by love or hatred of excitement, and by the fact that the bet is in money and not in goods and bads. Since it is universally agreed that money has a diminishing marginal utility, if money bets are to be used, it is evident that they should be for as small stakes as possible. But then again the measurement is spoiled by introducing the new factor of reluctance to bother about trifles.

Let us now discard the assumption that goods are additive and immediately measurable, and try to work out a system with as few assumptions as possible. To begin with we shall suppose, as before, that our subject has certain beliefs about everything; then he will act so that what he believes to be the total consequences of his action will be the best possible. If then we had the power of the Almighty, and could persuade our subject of our power, we could, by offering him options, discover how he placed in order of merit all possible courses of the world. In this way all possible worlds

would be put in an order of value, but we should have no definite way of
representing them by numbers. There would be no meaning in the assertion
that the difference in value between α and β was equal to that between γ and
δ. [Here and elsewhere we use Greek letters to represent the different possible totalities of events between which our subject chooses—the ultimate
organic unities.]

Suppose next that the subject is capable of doubt; then we could test his
degree of belief in different propositions by making him offers of the following kind. Would you rather have world α in any event; or world β if p is true,
and world γ if p is false? If, then, he were certain that p was true, he would
simply compare α and β and choose between them as if no conditions were
attached; but if he were doubtful his choice would not be decided so simply.
I propose to lay down axioms and definitions concerning the principles
governing choices of this kind. This is, of course, a very schematic version
of the situation in real life, but it is, I think easier to consider it in this form.

There is first a difficulty which must be dealt with; the propositions like p
in the above case which are used as conditions in the options offered may be
such that their truth or falsity is an object of desire to the subject. This will
be found to complicate the problem, and we have to assume that there are
propositions for which this is not the case, which we shall call ethically
neutral. More precisely an atomic proposition p is called ethically neutral if
two possible worlds differing only in regard to the truth of p are always of
equal value; and a non-atomic proposition p is called ethically neutral if all its
atomic truth-arguments[6] are ethically neutral.

We begin by defining belief of degree $\frac{1}{2}$ in an ethically neutral proposition.
The subject is said to have belief of degree $\frac{1}{2}$ in such a proposition p if he has
no preference between the options (1) α if p is true, β if p is false, and (2) α
if p is false, β if p is true, but has a preference between α and β simply. We
suppose by an axiom that if this is true of any one pair α, β it is true of all
such pairs.[7] This comes roughly to defining belief of degree $\frac{1}{2}$ as such a
degree of belief as leads to indifference between betting one way and betting
the other for the same stakes.

Belief of degree $\frac{1}{2}$ as thus defined can be used to measure values numerically in the following way. We have to explain what is meant by the difference
in value between α and β being equal to that between γ and δ; and we define
this to mean that, if p is an ethically neutral proposition believed to degree $\frac{1}{2}$,
the subject has no preference between the options (1) α if p is true, δ if p is
false, and (2) β if p is true, γ is p is false.

(6)   I assume here Wittgenstein's theory of propositions; it would probably
       be possible to give an equivalent definition in terms of any other.
(7) α and β must be supposed so far undefined as to be compatible with
       both p and not-p.

This defintion can form the basis of a system of measuring values in the following way.

Let us call any set of all worlds equally preferable to a given world a value: we suppose that if world α is preferable to β any world with the same value as α is preferable to any world with the same value as β and shall say that the value of α is greater than that of β. This relation "greater than" orders values in a series. We shall use α henceforth for both the world and its value.

AXIOMS.

(1) There is an ethically neutral proposition p believed to degree $\frac{1}{2}$.

(2) If p, q are such propositions and the option

α if p, δ if not-p is equivalent to β if p, γ if not-p

then α if q, δ if not-q is equivalent to β if q, γ if not-q.

Def. In the above case we say $\alpha\beta = \gamma\delta$.

THEOREMS.

If $\alpha\beta = \gamma\delta$,

then $\beta\alpha = \delta\gamma$, $\alpha\gamma = \beta\delta$, $\gamma\alpha = \delta\beta$.

(2a) If $\alpha\beta = \gamma\delta$, then $\alpha > \beta$ is equivalent to $\gamma > \delta$

and $\alpha = \beta$ is equivalent to $\gamma = \delta$.

(3) If option A is equivalent to option B and B to C then A to C.

THEOREM.

If $\alpha\beta = \gamma\delta$ and $\beta\eta = \zeta\gamma$,

then $\alpha\eta = \zeta\delta$.

(4) If $\alpha\beta = \gamma\delta$, $\gamma\delta = \eta\zeta$, then $\alpha\beta = \eta\zeta$.

(5) (α, β, γ). $E! (\imath x)(\alpha x = \beta\gamma)$.

(6) (α, β). $E! (\imath x)(\alpha x = x\beta)$.

(7) AXIOM OF CONTINUITY: Any progression has a limit (ordinal).

(8) AXIOM OF ARCHIMEDES.

These axioms enable the values to be correlated one-one with real numbers so that if α' corresponds to α, etc.

$$\alpha\beta = \gamma\delta. \equiv .\alpha' - \beta' = \gamma' - \delta'.$$

Henceforth we use α for the correlated real number α' also.

Having thus defined a way of measuring value we can now derive a way of measuring belief in general. If the option of α for certain is indifferent with that of β if p is true and γ if p is false,[8] we can define the subject's degree of

(8) Here β must include the truth of p, γ its falsity; p need no longer be ethically neutral. But we have to assume that there is a world with any assigned value in which p is true, and one in which p is false.

belief in p as the ratio of the difference between α and γ to that between β and γ; which we must suppose the same for all α's, β's and γ's that satisfy the conditions. This amounts roughly to defining the degree of belief in p by the odds at which the subject would bet on p, the bet being conducted in terms of differences of value as defined. The definition only applies to partial belief and does not include certain beliefs; for belief of degree 1 in p, α for certain is indifferent with α if p and any β if not-p.

We are also able to define a very useful new idea—"the degree of belief in p given q." This does not mean the degree of belief in "If p then q," or that in "p entails q," or that which the subject would have in p if he knew q, or that which he ought to have. It roughly expresses the odds at which he would now bet on p, the bet only to be valid if q is true. Such conditional bets were often made in the eighteenth century.

The degree of belief in p given q is measured thus. Suppose the subject indifferent between the options (1) α if q true, β if q false, (2) γ if p true and q true, δ if p false and q true, β if q false. Then the degree of his belief in p given q is the ratio of the difference between α and δ to that between γ and δ, which we must suppose the same for any $\alpha, \beta, \gamma, \delta$ which satisfy the given conditions. This is not the same as the degree to which he would believe p, if he believed q for certain; for knowledge of q might for psychological reasons profoundly alter his whole system of beliefs.

Each of our definitions has been accompanied by an axiom of consistency, and in so far as this is false, the notion of the corresponding degree of belief becomes invalid. This bears some analogy to the situation in regard to simultaneity discussed above.

I have not worked out the mathematical logic of this in detail, because this would, I think, be rather like working out to seven places of decimals a result only valid to two. My logic cannot be regarded as giving more than the sort of way it might work.

From these definitions and axioms it is possible to prove the fundamental laws of probable belief (degrees of belief lie between 0 and 1):

(1)   Degree of belief in p + degree of belief in $\bar{p}$ = 1.

(2)   Degree of belief in p given q + degree of belief in $\bar{p}$ given q = 1.

(3)   Degree of belief in (p and q) = degree of belief in p × degree of belief in q given p.

(4)   Degree of belief in (p and q) + degree of belief in (p and $\bar{q}$) = degree of belief in p.

The first two are immediate. (3) is proved as follows.

Let degree of belief in $p = x$, that in q given $p = y$.

Then ξ for certain $\equiv \xi + (1 - x)t$ if p true, $\xi - xt$ if p false, for any t.

$\xi + (1 - x)t$ if p true $\equiv$

$$\begin{cases} \xi + (1 - x)t + (1 - y)u \text{ if "}p \text{ and } q\text{" true,} \\ \xi + (1 - x)t - yu \text{ if } p \text{ true } q \text{ false; for any } u. \end{cases}$$

Choose u so that $\xi + (1 - x)t - yu = \xi - xt,$
   i.e. let $u = t/y$ $(y \neq 0)$.

Then ξ for certain $\equiv$

$$\begin{cases} \xi + (1 - x)t + (1 - y)t/y \text{ if } p \text{ and } q \text{ true} \\ \xi - xt \text{ otherwise,} \end{cases}$$

$\therefore$ degree of belief in "p and q" $= \dfrac{xt}{t + (1 - y)t/y} = xy.$ $(t \neq 0)$

If $y = 0$, take $t = 0$.

Then ξ for certain $\equiv \xi$ if p true, ξ if p false

                     $\equiv \xi + u$ if p true, q true; ξ if p false, q false;

                     ξ if p false

                     $\equiv \xi + u, pq$ true; ξ, pq false

$\therefore$ degree of belief in $pq = 0$.

(4)   follows from (2), (3) as follows:

Degree of belief in pq = that in $p \times$ that in q given p, by (3). Similarly degree
of belief in $p\overline{q}$ = that in $p \times$ that in $\overline{q}$ given p
$\therefore$ sum = degree of belief in p, by (2).

These are the laws of probability, which we have proved to be necessarily
true of any consistent set of degrees of belief. Any definite set of degrees of
belief which broke them would be inconsistent in the sense that it violated
the laws of preference between options, such as that preferability is a transitive asymmetrical relation, and that if α is preferable to β, β for certain cannot
be preferable to α if p, β if not-p. If anyone's mental condition violated
these laws, his choice would depend on the precise form in which the options
were offered him, which would be absurd. He could have a book made against
him by a cunning bettor and would then stand to lose in any event.

We find, therefore, that a precise account of the nature of partial belief
reveals that the laws of probability are laws of consistency, an extension to
partial beliefs of formal logic, the logic of consistency. They do not depend
for their meaning on any degree of belief in a proposition being uniquely
determined as the rational one; they merely distinguish those sets of beliefs
which obey them as consistent ones.

Having any definite degree of belief implies a certain measure of consistency, namely willingness to bet on a given proposition at the same odds for
any stake, the stakes being measured in terms of ultimate values. Having

degrees of belief obeying the laws of probability implies a further measure of consistency, namely such a consistency between the odds acceptable on different propositions as shall prevent a book being made against you.

Some concluding remarks on this section may not be out of place. First, it is based fundamentally on betting, but this will not seem unreasonable when it is seen that all our lives we are in a sense betting. Whenever we go to the station we are betting that a train will really run, and if we had not a sufficient degree of belief in this we should decline the bet and stay at home. The options God gives us are always conditional on our guessing whether a certain proposition is true. Secondly, it is based throughout on the idea of mathematical expectation; the dissatisfaction often felt with this idea is due mainly to the inaccurate measurement of goods. Clearly mathematical expectations in terms of money are not proper guides to conduct. It should be remembered, in judging my system, that in it value is actually defined by means of mathematical expectation in the case of beliefs of degree $\frac{1}{2}$, and so may be expected to be scaled suitably for the valid application of the mathematical expectation in the case of other degrees of belief also.

Thirdly, nothing has been said about degrees of belief when the number of alternatives is infinite. About this I have nothing useful to say, except that I doubt if the mind is capable of contemplating more than a finite number of alternatives. It can consider questions to which an infinite number of answers are possible, but in order to consider the answers it must lump them into a finite number of groups. The difficulty becomes practically relevant when discussing induction, but even then there seems to me no need to introduce it. We can discuss whether past experience gives a high probability to the sun's rising to-morrow without bothering about what probability it gives to the sun's rising each morning for evermore. For this reason I cannot but feel that Mr. Ritchie's discussion of the problem[9] is unsatisfactory; it is true that we can agree that inductive generalizations need have no finite probability, but particular expectations entertained on inductive grounds undoubtedly do have a high numerical probability in the

(9)   A. D. Ritchie, "Induction and Probability," *Mind* (1926), p. 318. "The conclusion of the foregoing discussion may be simply put. If the problem of induction be stated to be 'How can inductive generalizations acquire a large numerical probability?' then this is a pseudo-problem, because the answer is 'They cannot.' This answer is not, however, a denial of the validity of induction but is a direct consequence of the nature of probability. It still leaves untouched the real problem of induction which is 'How can the probability of an induction be increased?' and it leaves standing the whole of Keynes' discussion on this point."

minds of all of us. We all are more certain that the sun will rise tomorrow than that I shall not throw 12 with two dice first time, i.e. we have a belief of higher degree than 35/36 in it. If induction ever needs a logical justification it is in connection with the probability of an event like this.

4. The Logic of Consistency. We may agree that in some sense it is the business of logic to tell us what we ought to think; but the interpretation of this statement raises considerable difficulties. It may be said that we ought to think what is true, but in that sense we are told what to think by the whole of science and not merely by logic. Nor, in this sense, can any justification be found for partial belief; the ideally best thing is that we should have beliefs of degree 1 in all true propositions and beliefs of degree 0 in all false propositions. But this is too high a standard to expect of mortal men, and we must agree that some degree of doubt or even of error may be humanly speaking justified.

Many logicians, I suppose, would accept as an account of their science the opening words of Mr. Keynes' *Treatise on Probability*: "Part of our knowledge we obtain direct; and part by argument. The Theory of Probability is concerned with that part which we obtain by argument, and it treats of the different degrees in which the results so obtained are conclusive or inconclusive." Where Mr. Keynes says "the Theory of Probability," others would say Logic. It is held, that is to say, that our opinions can be divided into those we hold immediately as a result of perception or memory, and those which we derive from the former by argument. It is the business of Logic to accept the former class and criticize merely the derivation of the second class from them.

Logic as the science of argument and inference is traditionally and rightly divided into deductive and inductive; but the difference and relation between these two divisions of the subject can be conceived in extremely different ways. According to Mr. Keynes valid deductive and inductive arguments are fundamentally alike; both are justified by logical relations between premiss and conclusion which differ only in degree. This position, as I have already explained, I cannot accept. I do not see what these inconclusive logical relations can be or how they can justify partial beliefs. In the case of conclusive logical arguments I can accept the account of their validity which has been given by many authorities, and can be found substantially the same in Kant, De Morgan, Peirce and Wittgenstein. All these authors agree that the conclusion of a formally valid argument is contained in its premisses; that to deny the conclusion while accepting the premisses would be self-contradictory; that a formal deduction does not increase our knowledge, but only brings out clearly what we already know in another form; and that we are bound to accept its validity on pain of being inconsistent with ourselves. The logical relation which justifies the inference is that the sense or import of the conclusion is contained in that of the premisses.

But in the case of an inductive argument this does not happen in the least; it is impossible to represent it as resembling a deductive argument and merely weaker in degree; it is absurd to say that the sense of the conclusion is partially contained in that of the premises. We could accept the premises and utterly reject the conclusion without any sort of inconsistency or contradiction.

It seems to me, therefore, that we can divide arguments into two radically different kinds, which we can distinguish in the words of Peirce as (1) "explicative, analytic, or deductive" and (2) "ampliative, synthetic, or (loosely speaking) inductive."[10] Arguments of the second type are from an important point of view much closer to memories and perceptions than to deductive arguments. We can regard perception, memory and induction as the three fundamental ways of acquiring knowledge; deduction on the other hand is merely a method of arranging our knowledge and eliminating inconsistencies or contradictions.

Logic must then fall very definitely into two parts: (excluding analytic logic, the theory of terms and propositions) we have the lesser logic, which is the logic of consistency, or formal logic; and the larger logic, which is the logic of discovery, or inductive logic.

What we have now to observe is that this distinction in no way coincides with the distinction between certain and partial beliefs; we have seen that there is a theory of consistency in partial beliefs just as much as of consistency in certain beliefs, although for various reasons the former is not so important as the latter. The theory of probability is in fact a generalization of formal logic; but in the process of generalization one of the most important aspects of formal logic is destroyed. If p and $\bar{q}$ are inconsistent so that q follows logically from p, that p implies q is what is called by Wittgenstein a "tautology" and can be regarded as a degenerate case of a true proposition not involving the idea of consistency. This enables us to regard (not altogether correctly) formal logic including mathematics as an objective science consisting of objectively necessary propositions. It thus gives us not merely the ἀνάγκη λέγειν, that if we assert p we are bound in consistency to assert q also, but also the ἀνάγκη εἶναι, that if p is true, so must q be. But when we extend formal logic to include partial beliefs this direct objective interpretation is lost; if we believe pq to the extent of $\frac{1}{3}$, and $p\bar{q}$ to the extent of $\frac{1}{3}$, we are bound in consistency to believe $\bar{p}$ also to the extent of $\frac{1}{3}$. This is the ἀνάγκη λέγειν; but we cannot say that if pq is $\frac{1}{3}$ true and $p\bar{q}$ $\frac{1}{3}$ true, $\bar{p}$ also must be $\frac{1}{3}$ true, for such a statement would be sheer nonsense. There is no corresponding ἀνάγκη εἶναι. Hence, unlike the calculus of consistent full belief, the calculus of objective partial belief cannot be immediately interpreted as a body of objective tautology.

(10) C. S. Peirce, *Chance Love and Logic* (ed. M. Cohen), New York, 1923, p. 92.

This is, however, possible in a roundabout way; we saw at the beginning of this essay that the calculus of probabilities could be interpreted in terms of class-ratios; we have now found that it can also be interpreted as a calculus of consistent partial belief. It is natural, therefore, that we should expect some intimate connection between these two interpretations, some explanation of the possibility of applying the same mathematical calculus to two such different sets of phenomena. Nor is an explanation difficult to find; there are many connections between partial beliefs and frequencies. For instance, experienced frequencies often lead to corresponding partial beliefs, and partial beliefs lead to the expectation of corresponding frequencies in accordance with Bernouilli's Theorem. But neither of these is exactly the connection we want; a partial belief cannot in general be connected uniquely with any actual frequency, for the connection is always made by taking the proposition in question as an instance of a propositional function. What propositional function we choose is to some extent arbitrary and the corresponding frequency will vary considerably with our choice. The pretensions of some exponents of the frequency theory that partial belief means full belief in a frequency proposition cannot be sustained. But we found that the very idea of partial belief involves reference to a hypothetical or ideal frequency; supposing goods to be additive, belief of degree m/n is the sort of belief which leads to the action which would be best if repeated n times in m of which the proposition is true; or we can say more briefly that it is the kind of belief most appropriate to a number of hypothetical occasions otherwise identical in a proportion m/n of which the proposition in question is true. It is this connection between partial belief and frequency which enables us to use the calculus of frequencies as a calculus of consistent partial belief. And in a sense we may say that the two interpretations are the objective and subjective aspects of the same inner meaning, just as formal logic can be interpreted objectively as a body of tautology and subjectively as the laws of consistent thought.

We shall, I think, find that this view of the calculus of probability removes various difficulties that have hitherto been found perplexing. In the first place it gives us a clear justification for the axioms of the calculus, which on such a system as Mr. Keynes' is entirely wanting. For now it is easily seen that if partial beliefs are consistent they will obey these axioms, but it is utterly obscure why Mr. Keynes' mysterious logical relations should obey them.[11]

(11) It appears in Mr. Keynes' system as if the principal axioms—the laws of addition and multiplication—were nothing but definitions. This is merely a logical mistake; his definitions are formally invalid unless corresponding axioms are presupposed. Thus his definition of multiplication presupposes the law that if the probability of a given bh is equal to that of c given dk, and the probability of b given h is equal to that of d given k, then will the probabilities of ab given h and of cd given k be equal.

We should be so curiously ignorant of the instances of these relations, and so curiously knowledgeable about their general laws.

Secondly, the Principle of Indifference can now be altogether dispensed with; we do not regard it as belonging to formal logic to say what should be a man's expectation of drawing a white or a black ball from an urn; his original expectations may within the limits of consistency be any he likes; all we have to point out is that if he has certain expectations he is bound in consistency to have certain others. This is simply bringing probability into line with ordinary formal logic, which does not criticize premisses but merely declares that certain conclusions are the only ones consistent with them. To be able to turn the Principle of Indifference out of formal logic is a great advantage; for it is fairly clearly impossible to lay down purely logical conditions for its validity, as is attempted by Mr. Keynes. I do not want to discuss this question in detail, because it leads to hair-splitting and arbitrary distinctions which could be discussed for ever. But anyone who tries to decide by Mr. Keynes' methods what are the proper alternatives to regard as equally probable in molecular mechanics, e.g. in Gibbs' phase-space, will soon be convinced that it is a matter of physics rather than pure logic. By using the multiplication formula, as it is used in inverse probability, we can on Mr. Keynes' theory reduce all probabilities to quotients of *a priori* probabilities; it is therefore in regard to these latter that the Principle of Indifference is of primary importance; but here the question is obviously not one of formal logic. How can we on merely logical grounds divide the spectrum into equally probable bands?

A third difficulty which is removed by our theory is the one which is presented to Mr. Keynes' theory by the following case. I think I perceive or remember something but am not sure; this would seem to give me some ground for believing it, contrary to Mr. Keynes' theory, by which the degree of belief in it which it would be rational for me to have is that given by the probability relation between the proposition in question and the things I know for certain. He cannot justify a probable belief founded not on argument but on direct inspection. In our view there would be nothing contrary to formal logic in such a belief; whether it would be reasonable would depend on what I have called the larger logic which will be the subject of the next section; we shall there see that there is no objection to such a possibility, with which Mr. Keynes' method of justifying probable belief solely by relation to certain knowledge is quite unable to cope.

5. The Logic of Truth. The validity of the distinction between the logic of consistency and the logic of truth has been often disputed; it has been contended on the one hand that logical consistency is only a kind of factual consistency; that if a belief in p is inconsistent with one in q, that simply means that p and q are not both true, and that this is a necessary or logical

fact. I believe myself that this difficulty can be met by Wittgenstein's theory of tautology, according to which if a belief in p is inconsistent with one in q, that p and q are not both true is not a fact but a tautology. But I do not propose to discuss this question further here.

From the other side it is contended that formal logic or the logic of consistency is the whole of logic, and inductive logic either nonsense or part of natural science. This contention, which would I suppose be made by Wittgenstein, I feel more difficulty in meeting. But I think it would be a pity, out of deference to authority, to give up trying to say anything useful about induction.

Let us therefore go back to the general conception of logic as the science of rational thought. We found that the most generally accepted parts of logic, namely, formal logic, mathematics and the calculus of probabilities, are all concerned simply to ensure that our beliefs are not self-contradictory. We put before ourselves the standard of consistency and construct these elaborate rules to ensure its observance. But this is obviously not enough; we want our beliefs to be consistent not merely with one another but also with the facts:[12] nor is it even clear that consistency is always advantageous; it may well be better to be sometimes right than never right. Nor when we wish to be consistent are we always able to be: there are mathematical propositions whose truth or falsity cannot as yet be decided. Yet it may humanly speaking be right to entertain a certain degree of belief in them on inductive or other grounds: a logic which proposes to justify such a degree of belief must be prepared actually to go against formal logic; for to a formal truth formal logic can only assign a belief of degree 1. We could prove in Mr. Keynes' system that its probability is 1 on any evidence. This point seems to me to show particularly clearly that human logic or the logic of truth, which tells men how they should think is not merely independent of but sometimes actually incompatible with formal logic.

In spite of this nearly all philosophical thought about human logic and especially induction has tried to reduce it in some way to formal logic. Not that it is supposed, except by a very few, that consistency will of itself lead to truth; but consistency combined with observation and memory is frequently credited with this power.

Since an observation changes (in degree at least) my opinion about the fact observed, some of my degrees of belief after the observation are necessarily inconsistent with those I had before. We have therefore to explain how

(12) Cf. Kant: "Denn obgleich eine Erkenntnis der logischen Form völlig gemäss sein möchte, dass ist sich selbst nicht widerspräche, so kann sie doch noch immer dem Gegenstande widersprechen." *Kritik der reinen Vernunft,* First Edition, p. 59.

exactly the observation should modify my degrees of belief; obviously if p is the fact observed, my degree of belief in q after the observation should be equal to my degree of belief in q given p before, or by the multiplication law to the quotient of my degree of belief in pq by my degree of belief in p. When my degrees of belief change in this way we can say that they have been changed consistently by my observation.

By using this definition, or in Mr. Keynes' system simply by using the multiplication law, we can take my present degrees of belief, and by considering the totality of my observations, discover from what initial degrees of belief my present ones would have arisen by this process of consistent change. My present degrees of belief can then be considered logically justified if the corresponding initial degrees of belief are logically justified. But to ask what initial degrees of belief are justified, or in Mr. Keynes' system what are the absolutely *a priori* probabilities, seems to me a meaningless question; and even if it had a meaning I do not see how it could be answered.

If we actually applied this process to a human being, found out, that is to say, on what *a priori* probabilities his present opinions could be based, we should obviously find them to be ones determined by natural selection, with a general tendency to give higher probability to the simpler alternatives. But, as I say, I cannot see what could be meant by asking whether these degrees of belief were logically justified. Obviously the best thing would be to know for certain in advance what was true and what was false, and therefore if any one system of initial beliefs is to receive the philosopher's approbation it should be this one. But clearly this would not be accepted by thinkers of the school I am criticising. Another alternative is to apportion initial probabilities on the purely formal system expounded by Wittgenstein, but as this gives no justification for induction it cannot give us the human logic which we are looking for.

Let us therefore try to get an idea of a human logic which shall not attempt to be reducible to formal logic. Logic, we may agree, is concerned not with what men actually believe, but what they ought to believe, or what it would be reasonable to believe. What then, we must ask, is meant by saying that it is reasonable for a man to have such and such a degree of belief in a proposition? Let us consider possible alternatives.

First, it sometimes means something explicable in terms of formal logic: this possibility for reasons already explained we may dismiss. Secondly, it sometimes means simply that were I in his place (and not e.g. drunk) I should have such a degree of belief. Thirdly, it sometimes means that if his mind worked according to certain rules, which we may roughly call "scientific method," he would have such a degree of belief. But fourthly it need mean none of these things; for men have not always believed in scientific method, and just as we ask "But am I necessarily reasonable," we can also ask "But is the scientist necessarily reasonable?" In this ultimate meaning it seems to me that we can identify reasonable opinion with the opinion of an ideal person in similar circumstances. What, however, would the ideal person's opinion be? As has previously been remarked, the highest ideal would be always to have a true opinion and be certain of it; but this ideal is more suited to God than to man.[13]

We have therefore to consider the human mind and what is the most we can ask of it.[14] The human mind works essentially according to general rules or habits; a process of thought not proceeding according to some rule would simply be a random sequence of ideas; whenever we infer A from B we do so in virtue of some relation between them. We can therefore state the problem of the ideal as "What habits in a general sense would it be best for the human mind to have?" This is a large and vague question which could hardly be answered unless the possibilities were first limited by a fairly definite conception of human nature. We could imagine some very useful habits unlike those

(13) [Earlier draft of matter of preceding paragraph in some ways better.— F.P.R.

What is meant by saying that a degree of belief is reasonable? First and often that it is what I should entertain if I had the opinions of the person in question at the time but was otherwise as I am now, e.g. not drunk. But sometimes we go beyond this and ask: "Am I reasonable?" This may mean, do I conform to certain enumerable standards which we call scientific method, and which we value on account of those who practise them and the success they achieve. In this sense to be reasonable means to think like a scientist, or to be guided only by ratiocination and induction or something of the sort (i.e. reasonable means reflective). Thirdly, we may go to the root of why we admire the scientist and criticize not primarily an individual opinion but a mental habit as being conducive or otherwise to the discovery of truth or to entertaining such degrees of belief as will be most useful. (To include habits of doubt or partial belief.) Then we can criticize an opinion according to the habit which produced it. This is clearly right because it all depends on this habit; it would not be reasonable to get the right conclusion to a syllogism by remembering vaguely that you leave out a term which is common to both premisses.

We use reasonable in sense 1 when we say of an argument of a scientist this does not seem to me reasonable; in sense 2 when we *contrast* reason and superstition or instinct; in sense 3 when we *estimate* the value of new methods of thought such as soothsaying.]

(14) What follows to the end of the section is almost entirely based on the writings of C. S. Peirce. [Especially his "Illustrations of the Logic of Science," *Popular Science Monthly* (1877 and 1878), reprinted in *Chance Love and Logic*, 1923.]

possessed by any men. (It must be explained that I use habit in the most general possible sense to mean simply rule or law of behavior, including instinct: I do not wish to distinguish acquired rules or habits in the narrow sense from innate rules or instincts, but propose to call them all habits alike.) A completely general criticism of the human mind is therefore bound to be vague and futile, but something useful can be said if we limit the subject in the following way.

Let us take a habit of forming opinion in a certain way; e.g. the habit of proceeding from the opinion that a toadstool is yellow to the opinion that it is unwholesome. Then we can accept the fact that the person has a habit of this sort, and ask merely what degree of opinion that the toadstool is unwholesome it would be best for him to entertain when he sees it; i.e. granting that he is going to think always in the same way about all yellow toadstools, we can ask what degree of confidence it would be best for him to have that they are unwholesome. And the answer is that it will in general be best for his degree of belief that a yellow toadstool is unwholesome to be equal to the proportion of yellow toadstools which are in fact unwholesome. (This follows from the meaning of degree of belief.) This conclusion is necessarily vague in regard to the spatio-temporal range of toadstools which it includes, but hardly vaguer than the question which it answers. (Cf. density at a point of gas composed of molecules.)

Let us put it in another way: whenever I make an inference I do so according to some rule or habit. An inference is not completely given when we are given the premiss and conclusion; we require also to be given the relation between them in virtue of which the inference is made. The mind works by general laws; therefore if it infers q from p, this will generally be because q is an instance of a function ϕx and p the corresponding instance of a function ψx such that the mind would always infer ϕx from ψx. When therefore we criticize not opinions but the processes by which they are formed, the rule of the inference determines for us a range to which the frequency theory can be applied. The rule of the inference may be narrow, as when seeing lightning I expect thunder, or wide, as when considering 99 instances of a generalization which I have observed to be true I conclude that the 100th is true also. In the first case the habit which determines the process is "After lightning expect thunder"; the degree of expectation which it would be best for this habit to produce is equal to the proportion of cases of lightning which are actually followed by thunder. In the second case the habit is the more general one of inferring from 99 observed instances of a certain sort of generalization that the 100th instance is true also; the degree of belief it would be best for this habit to produce is equal to the proportion of all cases of 99 instances of a generalization being true, in which the 100th is true also.

Thus given a single opinion, we can only praise or blame it on the ground of truth or falsity: given a habit of a certain form, we can praise or blame it accordingly as the degree of belief it produces is near or far from the actual proportion in which the habit leads to truth. We can then praise or blame opinions derivatively from our praise or blame of the habits that produce them.

This account can be applied not only to habits of inference but also to habits of observation and memory; when we have a certain feeling in connection with an image we think the image represents something which actually happened to us, but we may not be sure about it; the degree of direct confidence in our memory varies. If we ask what is the best degree of confidence to place in a certain specific memory feeling, the answer must depend on how often when that feeling occurs the event whose image it attaches to has actually taken place.

Among the habits of the human mind a position of peculiar importance is occupied by induction. Since the time of Hume a great deal has been written about the justification for inductive inference. Hume showed that it could not be reduced to deductive inference or justified by formal logic. So far as it goes his demonstration seems to me final; and the suggestion of Mr. Keynes that it can be got round by regarding induction as a form of probable inference cannot in my view be maintained. But to suppose that the situation which results from this is a scandal to philosophy is, I think, a mistake.

We are all convinced by inductive arguments, and our conviction is reasonable because the world is so constituted that inductive arguments lead on the whole to true opinions. We are not, therefore, able to help trusting induction, nor if we could help it do we see any reason why we should, because we believe it to be a reliable process. It is true that if any one has not the habit of induction, we cannot prove to him that he is wrong; but there is nothing peculiar in that. If a man doubts his memory or his perception we cannot prove to him that they are trustworthy; to ask for such a thing to be proved is to cry for the moon, and the same is true of induction. It is one of the ultimate sources of knowledge just as memory is: no one regards it as a scandal to philosophy that there is no proof that the world did not begin two minutes ago and that all our memories are not illusory.

We all agree that a man who did not make inductions would be unreasonable: the question is only what this means. In my view it does not mean that the man would in any way sin against formal logic or formal probability; but that he had not got a very useful habit, without which he would be very much worse off, in the sense of being much less likely[15] to have true opinions.

(15) "Likely" here simply means that I am not sure of this, but only have a certain degree of belief in it.

This is a kind of pragmatism: we judge mental habits by whether they work, i.e. whether the opinions they lead to are for the most part true, or more often true than those which alternative habits would lead to.

Induction is such a useful habit, and so to adopt it is reasonable. All that philosophy can do is to analyse it, determine the degree of its utility, and find on what characteristics of nature this depends. An indispensable means for investigating these problems is induction itself, without which we should be helpless. In this circle lies nothing vicious. It is only through memory that we can determine the degree of accuracy of memory; for if we make experiments to determine this effect, they will be useless unless we remember them.

Let us consider in the light of the preceding discussion what sort of subject is inductive or human logic—the logic of truth. Its business is to consider methods of thought, and discover what degree of confidence should be placed in them, i.e. in what proportion of cases they lead to truth. In this investigation it can only be distinguished from the natural sciences by the greater generality of its problems. It has to consider the relative validity of different types of scientific procedure, such as the search for a causal law by Mill's Methods, and the modern mathematical methods like the *a priori* arguments used in discovering the Theory of Relativity. The proper plan of such a subject is to be found in Mill[16]; I do not mean the details of his Methods or even his use of the Law of Causality. But his way of treating the subject as a body of inductions about inductions, the Law of Causality governing lesser laws and being itself proved by induction by simple enumeration. The different scientific methods that can be used are in the last resort judged by induction by simple enumeration; we choose the simplest law that fits the facts, but unless we found that laws so obtained also fitted facts other than those they were made to fit, we should discard this procedure for some other.

(16) Cf. also the account of "general rules" in the Chapter "Of Unphilosophical Probability" in Hume's *Treatise*.

BRUNO DE FINETTI

Foresight: Its Logical Laws, Its Subjective Sources
(1937)

This article appeared in the *Annales de l'Institut Henri Poincaré*, Vol. 7 (1937). Reprinted by permission of the author and the publisher, Gauthier-Villars. Translated by Henry E. Kyburg, Jr.

The translator wishes to express his gratitude for their time and advice on many matters to Professor L. J. Savage, Professor Bruno de Finetti, and Miss Caroline Clauser. The translation has benefited greatly from their suggestions. He particularly wishes to express his gratitude to de Finetti for suggesting the following changes.

Words: The word "equivalent" of the original has been translated throughout as "exchangeable." The original term (used also by Khinchin) and even the term "symmetric" (used by Savage and Hewitt) appear to admit ambiguity. The word "exchangeable," proposed by Fréchet, seems expressive and unambiguous and has been adopted and recommended by most authors, including de Finetti.

The word "subjectiv" was used ambiguously in the original paper, both in the sense of "subjective" or "personal," as in "subjective probability," and in the sense of "subjectivistic," as in "the subjectivistic theory of probability," where "subjectiv" does not mean subjective (personal, private) at all. The distinction between the two concepts is made throughout the translation; the word "subjectivist" is reserved to mean "one who holds a subjectivistic theory."

"Cohérence" has been translated "coherent" following the usage of Shimony, Kemeny, and others. "Consistency" is used by some English and American authors, and is perfectly acceptable to de Finetti, but it is ambiguous (from the logician's point of view) because, applied to beliefs, it has another very precise and explicit meaning in formal logic. As the words are used in this translation, to say that a body of beliefs is "consistent" is to say (as in logic) that it contains no two beliefs that are contradictory. To say that in addition the body of beliefs is "coherent" is to say that the *degrees* of belief satisfy certain further conditions.

"Nombre aléatoire" has been translated as "random quantity." Although the phrase "random variable" is far more familiar to English-speaking mathematicians and philosophers, there are excellent reasons, as de Finetti points out, for making this substitution. I shall quote two of these reasons from de Finetti's correspondence. The first reason is that emphasized repeatedly in connection with the word "event." "While frequentists speak of an event as something admitting repeated 'trials,' for those who take a subjectivistic (or logical) view of probability, any trial is a different 'event.' Likewise, for frequentists, a random variable X is something assuming different values in repeated 'trials,' and only with this interpretation is the word 'variable' proper. For me any single trial gives a *random quantity*; there is nothing *variable*: the value is univocally indicated; it is only *unknown*; there is only *uncertainty* (for me, for somebody) about the unique value it will exhibit." The second objection de Finetti raises to the phrase "random variable" is one that is quite independent of any particular point of view with respect to probability. "Even with the statistical conception of probability, it is unjustifiably asymmetric to speak of random points, random functions, random vectors, etc., and of random variables when the 'variable' is a number or quantity; it would be consistent to say 'random variable' *always*, specifying, if necessary, 'random variable numbers,' 'random variable points,' 'random variable vectors,' 'random variable functions,' etc., as particular kinds of random variables."

"Loi" is used in the text both in the sense of "*theorem*" (as in "the law of large numbers") and in the sense of "distribution" (as in "normal law"). This is conventional French usage, and to some extent English and American usage has followed the French in this respect. But de Finetti himself now avoids the ambiguity by reserving the word "law" for the first sense (theoremhood) only, and by introducing the term "distribution" in a general sense to serve the function of the word "law" in its second sense. "Distribution" in this general sense may refer to specific distribution functions (as in "normal distribution"), the additive function of events P(E), or distributions that are not indicated by particular functions at all. I have attempted, with de Finetti's advice and suggestions, to introduce this distinction in translation.

Notation: The original notation has been followed closely, with the single exception of that for the "conditional event," E given A, which is written in the (currently) usual way, $E|A$. In the original this is written $\frac{E}{A}$. I have also substituted the conventional "∨" for the original "+" in forming the expression denoting the disjunction of two events.

Footnotes: Professor de Finetti has very kindly provided us with new notes that give some indication of the changes that have occurred in his thinking since he wrote this paper, or which clarify points which have, since the original writing, appeared to need clarification. These new notes are indicated by italic letters; the numbered footnotes appeared in the original work.

Translator's Note to the Second Edition: Professor de Finetti has kindly pointed out a number of errors in the translation as it appeared in the first edition, both in personal correspondence and in the Introduction to *Probability, Induction, and Statistics*, p. xvii. These have been corrected in the present edition.

FOREWORD

·In the lectures which I had the honor to give at the Institut Henri Poincaré the second, third, eighth, ninth, and tenth of May 1935, the text of which is reproduced in the pages that follow, I attempted to give a general view of two subjects which particularly interest me, and to clarify the delicate relationship that unites them. There is the question, on the one hand, of the definition of probability (which I consider a purely subjective entity) and of the meaning of its laws, and, on the other hand, of the concepts and of the theory of "exchangeable" events and random quantities; the link between the two subjects lies in the fact that the latter theory provides the solution of the problem of inductive reasoning for the most typical case, according to the subjectivistic conception of probability (and thus clarifies, in general, the way in which the problem of induction is posed). Besides, even if this were not so, that is to say, even if the subjective point of view which we have adopted were not accepted, this theory would have no less validity and would still be an interesting chapter in the theory of probability.

The exposition is divided into six chapters, of which the first two deal with the first question, the following two with the second, and of which the last two examine the conclusions that can be drawn. The majority of the questions treated here have been dealt with, sometimes in detail, sometimes briefly, but always in a fragmentary way,[1] in my earlier works. Among these, those which treat questions studied or touched upon in these lectures are indicated in the bibliography.[2]

For more complete details concerning the material in each of these chapters, I refer the reader to the following publications.

Chapter I. The logic of the probable: [26], [34].
    II. The evaluation of probability: [49], [63], [70].
    III. Exchangeable events: [29], [40].
    IV. Exchangeable random quantities: [46], [47], [48].
    V. Reflections on the notion of exchangeability: [51], [62].
    VI. Observation and prediction: [32], [36], [62].

(1)   A more complete statement of my point of view, in the form of a purely critical and philosophical essay, without formulas, is to be found in [32].

(2)   See page 115; the numbers in boldface type refer always to this list (roman numerals for the works of other authors; arabic numerals for my own, arranged by general chronological order).

Each of these chapters constitutes one of the five lectures,[3] with the exception of Chapters IV and V, which correspond to the fourth, in which the text has been amplified in order to clarify the notion used there of integration in function space. The text of the other lectures has not undergone any essential modifications beyond a few improvements, for example, at the beginning of Chapter III, where, for greater clarity, the text has been completely revised. For these revisions, I have profited from the valuable advice of MM. Fréchet and Darmois, who consented to help with the lectures, and of M. Castelnuovo, who read the manuscript and its successive modifications several times; the editing of the text has been reviewed by my colleague M. V. Carmona and by M. Al. Proca, who suggested to me a number of stylistic changes. For their kind help I wish to express here my sincere appreciation. Finally, I cannot end these remarks without again thanking the director and the members of the governing committee of the Institut Henri Poincaré for the great honor they have done me by inviting me to give these lectures in Paris.

Trieste, December 19, 1936

(3)   Their titles are those of the six chapters, with the exception of Chapter V.

INTRODUCTION

Henri Poincaré, the immortal scientist whose name this institute honors, and who brought to life with his ingenious ideas so many branches of mathematics, is without doubt also the thinker who attributed the greatest domain of application to the theory of probability and gave it a completely essential role in scientific philosophy. "Predictions," he said, "can only be probable. However solidly founded a prediction may appear to us, we are never absolutely sure that experience will not refute it." The calculus of probability rests on "an obscure instinct, which we cannot do without; without it science would be impossible, without it we could neither discover a law nor apply it." "On this account all the sciences would be but unconscious applications of the calculus of probability; to condemn this calculus would be to condemn science entirely."[1]

Thus questions of principle relating to the significance and value of probability cease to be isolated in a particular branch of mathematics and take on the importance of fundamental epistemological problems.

Such questions evidently admit as many different answers as there are different philosophical attitudes; to give one answer does not mean to say something that can convince and satisfy everybody, but familiarity with one particular point of view can nevertheless be interesting and useful even to those who are not able to share it. The point of view I have the honor of presenting here may be considered the extreme of subjectivistic solutions; the link uniting the diverse researches that I propose to summarize is in fact the principal common goal which is pursued in all of them, beyond other, more immediate and concrete objectives; this goal is that of bringing into the framework of the subjectivistic conception and of explaining even the problems that seem to refute it and are currently invoked against it. The aim of the first lecture will be to show how the logical laws of the theory of probability can be rigorously established within the subjectivistic point of view; in the others it will be seen how, while refusing to admit the existence of an objective meaning and value for probabilities, one can get a clear idea of the reasons, themselves subjective, for which in a host of problems the subjective judgments of diverse normal individuals not only do not differ essentially from each other, but even coincide exactly. The simplest cases will be the subject of the second lecture; the following lectures will be devoted to the most delicate question of this study: that of understanding the subjectivistic

(1)   [XXVIII], p. 183, 186.

explanation of the use we make of the results of observation, of past experience, in our predictions of the future.

This point of view is only one of the possible points of view, but I would not be completely honest if I did not add that it is the only one that is not in conflict with the logical demands of my mind. If I do not wish to conclude from this that it is "true," it is because I know very well that, as paradoxical as it seems, nothing is more subjective and personal than this "instinct of that which is logical" which each mathematician has, when it comes to the matter of applying it to questions of principle.

CHAPTER I

The Logic of the Probable

Let us consider the notion of probability as it is conceived by all of us in everyday life. Let us consider a well-defined event and suppose that we do not know in advance whether it will occur or not; the doubt about its occurrence to which we are subject lends itself to comparison, and, consequently, to gradation. If we acknowledge only, first, that one uncertain event can only appear to us (a) equally probable, (b) more probable, or (c) less probable than another; second, that an uncertain event always seems to us more probable than an impossible event and less probable than a necessary event; and finally, third, that when we judge an event E' more probable than an event E, which is itself judged more probable than an event E'', the event E' can only appear more probable than E'' (transitive property), it will suffice to add to these three evidently trivial axioms a fourth, itself of a purely qualitative nature, in order to construct rigorously the whole theory of probability. This fourth axiom tells us that inequalities are preserved in logical sums: if E is incompatible with E_1 and with E_2, then $E_1 \vee E$ will be more or less probable than $E_2 \vee E$, or they will be equally probable, according to whether E_1 is more or less probable than E_2, or they are equally probable. More generally, it may be deduced from this[2] that two inequalities, such as

$$E_1 \text{ is more probable than } E_2,$$

$$E_1' \text{ is more probable than } E_2',$$

can be added to give

$$E_1 \vee E_1' \text{ is more probable than } E_2 \vee E_2',$$

(2)   See [34], p. 321, note 1.

provided that the events added are incompatible with each other (E_1 with E_1', E_2 with E_2'). It can then be shown that when we have events for which we know a subdivision into possible cases that we judge to be equally probable, the comparison between their probabilities can be reduced to the purely arithmetic comparison of the ratio between the number of favorable cases and the number of possible cases (not because the judgment then has an objective value, but because everything substantial and thus subjective is already included in the judgment that the cases constituting the division are equally probable). This ratio can then be chosen as the appropriate index to measure a probability, and applied in general, even in cases other than those in which one can effectively employ the criterion that governs us there. In these other cases one can evaluate this index by comparison: it will be in fact a number, uniquely determined, such that to numbers greater or less than that number will correspond events respectively more probable or less probable than the event considered. Thus, while starting out from a purely qualitative system of axioms, one arrives at a quantitative measure of probability, and then at the theorem of total probability which permits the construction of the whole calculus of probabilities (for conditional probabilities, however, it is necessary to introduce a fifth axiom: see note 8, p. 69).

One can, however, also give a direct, quantitative, numerical definition of the degree of probability attributed by a given individual to a given event, in such a fashion that the whole theory of probability can be deduced immediately from a very natural condition having an obvious meaning. It is a question simply of making mathematically precise the trivial and obvious idea that the degree of probability attributed by an individual to a given event is revealed by the conditions under which he would be disposed to bet on that event.[3] The axiomatization whose general outline we have just indicated above has the advantage of permitting a deeper and more detailed analysis, of starting out with only qualitative notions, and of eliminating the notion of "money," foreign to the question of probability, but which is required to talk of stakes; however, once it has been shown that one can overcome the distrust that is born of the somewhat too concrete and perhaps artificial nature of the definition based on bets, the second procedure is preferable, above all for its clarity.

(3) Bertrand ([1], p. 24) beginning with this observation, gave several examples of subjective probabilities, but only for the purpose of contrasting them with "objective probabilities." The subjectivistic theory has been developed according to the scheme of bets in the exposition (Chap. I and II) in my first paper of 1928 on this subject. This was not published in its original form, but was summarized or partially developed in [27], [34], [35], etc.

Let us suppose that an individual is obliged to evaluate the rate p at which he would be ready to exchange the possession of an arbitrary sum S (positive or negative) dependent on the occurrence of a given event E, for the possession of the sum pS; we will say by definition that this number p is the measure of the degree of probability attributed by the individual considered to the event E, or, more simply, that p is the probability of E (according to the individual considered; this specification can be implicit if there is no ambiguity.)[a]

Let us further specify that, in the terminology that I believe is suitable to follow, an event is always a singular fact; if one has to consider several trials, we will never say "trials of the same event" but "trials of the same phenomenon" and each "trial" will be one "event." The point is obviously not the choice of terms: it is a question of making precise that, according to us, one has no right to speak of the "probability of an event" if one understands by "event" that which we have called a "phenomenon"; one can only do this if it is a question of one specific "trial."[4]

This being granted, once an individual has evaluated the probabilities of certain events, two cases can present themselves: either it is possible to bet

(a) Such a formulation could better, like Ramsey's, deal with expected *utilities*; I did not know of Ramsey's work before 1937, but I was aware of the difficulty of money bets. I preferred to get around it by considering sufficiently small stakes, rather than to build up a complex theory to deal with it. I do not remember whether I failed to mention this limitation to small amounts inadvertently or for some reason, for instance considering the difficulty overcome in the artificial situation of compulsory choice.

Another shortcoming of the definition—or of the device for making it operational—is the possibility that people accepting bets against our individual have better information than he has (or know the outcome of the event considered). This would bring us to game-theoretic situations.

Of course, a device is always imperfect, and we must be content with an idealization. A better device (in this regard) is that mentioned in B. de Finetti and L. J. Savage, "Sul modo di scegliere le probabilità iniziali," *Biblioteca del Metron*, S. C. Vol. 1, pp. 81-147 (English summary pp. 148-151), and with some more detail in B. de Finetti, "Does it make sense to speak of 'good probability appraisers'?" *The Scientist Speculates: An anthology of partly-baked ideas*, Gen. Ed. I. J. Good, Heinemann, London, 1962. This device will be fully presented by the same authors in a paper in preparation.

(4) This same point of view has been taken by von Kries [XIX]; see [65], [70], and, for the contrary point of view, see [XXV].

with him in such a way as to be assured of gaining, or else this possibility does not exist. In the first case one clearly should say that the evaluation of the probabilities given by this individual contains an incoherence, an intrinsic contradiction; in the other case we will say that the individual is coherent.[b] It is precisely this condition of coherence which constitutes the sole principle from which one can deduce the whole calculus of probability: this calculus then appears as a set of rules to which the subjective evaluation of probability of various events by the same individual ought to conform if there is not to be a fundamental contradiction among them.

Let us see how to demonstrate, on this view, the theorem of total probability: it is an important result in itself, and also will clarify the point of view followed. Let $E_1, E_2, \ldots, E_n$ be incompatible events, of which one (and one only) must occur (we shall say: a *complete* class of incompatible events), and let $p_1, p_2, \ldots, p_n$ be their probabilities evaluated by a given individual; if one fixes the stakes (positive or negative) $S_1, S_2, \ldots, S_n$, the gains in the n possible cases will be the difference between the stake of the bet won and the sum of the n paid outlays.

$$G_h = S_h - \sum_1^n p_i S_i$$

By considering the S_h as unknowns, one obtains a system of linear equations with the determinant

$$\begin{vmatrix} 1 - p_1 & -p_2 & \cdots & -p_n \\ -p_1 & 1 - p_2 & \cdots & -p_n \\ \cdots & \cdots & \cdots & \cdots \\ -p_1 & -p_2 & \cdots & 1 - p_n \end{vmatrix} = 1 - (p_1 + p_2 + \cdots + p_n);$$

if this determinant is not zero, one can fix the S_h in such a way that the G_h have arbitrary values, in particular, all positive, contrary to the condition of

(b)   To speak of coherent or incoherent (consistent or inconsistent) individuals has been interpreted as a criticism of people who do not accept a specific behavior rule. Needless to say, this is meant only as a technical distinction. At any rate, it is better to speak of coherence (consistency) of probability evaluations rather than of individuals, not only to avoid this charge, but because the notion belongs strictly to the evaluations and only indirectly to the individuals. Of course, an individual may make mistakes sometimes, often without meriting contempt.

coherence; consequently coherence obliges us to impose the condition $p_1 + p_2 \cdots + p_n = 1.^c$ This necessary condition for coherence is also sufficient because, if it is satisfied, one has identically (whatever be the stakes S_h)

$$\sum_{1}^{n}{}_h\, p_h\, G_h = 0$$

and the G_h can never, in consequence, all be positive.

Thus one has the theorem of total probabilities in the following form: *in a complete class of incompatible events, the sum of the probabilities must be equal to 1.* The more general form, *the probability of the logical sum of n incompatible events is the sum of their probabilities,* is only an immediate corollary.

However, we have added that the condition is also sufficient; it is useful to make the sense of this assertion a little clearer, for in a concrete case one can throw into clear relief the distinction, fundamental from this point of view, between the logic of the probable and judgments of probability. In saying that the condition is sufficient, we mean that, a complete class of incompatible events $E_1, E_2, \ldots, E_n$ being given, all the assignments of probability that attribute to $p_1, p_2, \ldots, p_n$ any values whatever, which are non-negative and have a sum equal to unity, are admissible assignments: each of these evaluations corresponds to a coherent opinion, to an opinion legitimate in itself, and every individual is free to adopt that one of these opinions which he prefers, or, to put it more plainly, that which he *feels*. The best example is that of a championship where the spectator attributes to each team a greater or smaller probability of winning according to his own judgment; the theory cannot reject *a priori* any of these judgments unless the sum of the probabilities attributed to each team is not equal to unity. This arbitrariness, which any one would admit in the above case, exists also, according to the conception which we are maintaining, in all other domains, including those more or less vaguely defined domains in which the various objective conceptions are asserted to be valid.

Because of this arbitrariness, the subject of the calculus of probabilities is no longer a single function $P(E)$ of events E, that is to say, their probability considered as something objectively determined, but the set of all functions $P(E)$ corresponding to admissible opinions. And when a *calculation*

(c)   Of course the proof might have been presented in an easier form by considering simply the case of $S_1 = S_2 = \cdots = S_n = S$ (as I did in earlier papers). On this occasion I preferred a different proof which perhaps gives deeper insight.

of the probability $P(E)$ of an event E is wanted, the statement of the problem is to be made precise in this sense: calculate the value that one is obliged to attribute to the event E if one wants to remain in the domain of coherence, after having assigned definite probabilities to the events constituting a certain class $\mathcal{E}$. Mathematically the function P is adopted over the set $\mathcal{E}$, and one asks what unique value or what set of values can be attributed to $P(E)$ without this extension of P making an incoherence appear.

It is interesting to pose the following general question: what are the events E for which the probability is determined by the knowledge of the probabilities attributed to the events of a given class $\mathcal{E}$? We are thus led to introduce the notion (which I believe novel) of "linearly independent events" [26]. Let $E_1, E_2, \ldots, E_n$ be the events of $\mathcal{E}$. Of these n events some will occur, others will not; there being 2^n subclasses of a class of n elements (including the whole class $\mathcal{E}$ and the empty class), there will be at most 2^n possible cases $C_1, C_2, \ldots, C_s$ $(s \leqslant 2^n)$ which we call, after Boole, "constituents." ("At most," since a certain number of combinations may be impossible.)[5] Formally, the C_h are the events obtained by starting with the logical product $E_1 \cdot E_2 \cdot \ldots \cdot E_n$ and replacing any group of E_i by the contrary events (negations) $\sim E_i$ (or, in brief notation, $\overline{E}_i$). The constituents form a complete class of incompatible events; the E_i are logical sums of constituents, and the events which are the sums of constituents are the only events logically dependent on the E_i, that is, such that one can always say whether they are true or false when one knows, for each event $E_1, \ldots, E_n$, if it is true or false.

To give the probability of an event E_i means to give the sum of the probabilities of its constituents

$$c_{i_1} + c_{i_2} + \cdots + c_{i_h} = p_i;$$

the probabilities of $E_1, \ldots, E_n$ being fixed, one obtains n equations of this type, which form, with the equation $c_1 + c_2 + \cdots + c_s = 1$, a system of $n + 1$ linear equations relating the probabilities c_h of the constituents. It may be seen that, E being an event logically dependent on $E_1, \ldots, E_n$, and thus a logical sum of constituents $E = C_{h_1} \vee C_{h_2} \vee \cdots \vee C_{h_k}$, its probability

$$p = c_{h_1} + c_{h_2} + \cdots + c_{h_k}$$

is uniquely determined when this equation is linearly dependent on the preceding system of equations. Observe that this fact does not depend on the function P, but only on the class $\mathcal{E}$ and the event E and can be expressed

(5)   These notions are applied to the calculus of probability in Medolaghi [XXIV].

by saying that E is *linearly dependent on* $\mathscr{E}$, or—what comes to the same thing if the E_i are linearly independent—that E_1, E_2, ..., E_n and E are linearly related among themselves.

The notion of linear independence thus defined for events is perfectly analogous to the well-known geometrical notion, and enjoys the same properties; instead of this fact being demonstrated directly, it can quickly be made obvious by introducing a geometrical representation which makes a point correspond to each event, and the notion of geometrical "linear independence" correspond to the notion of logical "linear independence." The representation is as follows: the constituents C_h are represented by the Apexes A_h of a simplex in a space of $s - 1$ dimensions, the event which is the sum of k constituents by the center of gravity of the k corresponding apexes given a mass k, and finally, the certain event (the logical sum of all the s constituents) by the center 0 of the simplex, given a mass s.

This geometric representation allows us to characterize by means of a model the set of all possible assignments of probability. We have seen that a probability function $\mathbf{P}(E)$ is completely determined when one gives the relative values of the constituents, $c_1 = \mathbf{P}(C_1)$, $c_2 = \mathbf{P}(C_2)$, ..., $c_s = \mathbf{P}(C_s)$, values which must be non-negative and have a sum equal to unity. Let us now consider the linear function f which takes the values c_h on the apexes A_h; at the point A, the center of gravity of A_{h_1}, A_{h_2}, ..., A_{h_k}, it obviously takes the value $f(A) = (1/k)(c_{h_1} + c_{h_2} + \cdots + c_{h_k})$, while the probability $\mathbf{P}(E)$ of the event E, the logical sum of the constituents C_{h_1}, C_{h_2}, ..., C_{h_k} will be $c_{h_1} + c_{h_2} + \cdots + c_{h_k}$. We have, then, in general, $\mathbf{P}(E) = k \cdot f(A)$: the probability of an event E is the value of f at its representative point A, multiplied by the mass k; one could say that it is given as the value of f for the point A endowed with a mass k, writing $\mathbf{P}(E) = f(k \cdot A)$.[d] The center 0 corresponding to the certain event, one has in particular $1 = f(s \cdot 0) = s \cdot f(0)$, that is, $f(0) = (1/s)$.

It is immediately seen that the possible assignments of probability correspond to all the linear functions of the space that are non-negative on the simplex and have the value $1/s$ at the origin; such a function f being characterized by the hyperplane $f = 0$, assignments of probability correspond biunivocally to the hyperplanes which do not cut the simplex. It may be seen that the probability $\mathbf{P}(E) = f(k \cdot A)$ is the moment of the given mass point kA (distance $\times$ mass) relative the the hyperplane $f = 0$ (taking as unity the moment of s0). If, in particular, the s constituents are equally probable, the hyperplane goes to infinity.

(d)   The notion of "weighted point," or "geometrical formation of the first kind," belongs to the geometrical approach and notations of Grassmann-Peano, to which the Italian school of vector calculus adheres.

By giving the value that it takes on a certain group of points, a linear function f is defined for all those points linearly dependent on them, but it remains undetermined for linearly independent points: the comparison with the above definition of linearly dependent events thus shows, as we have said, that the linear dependence and independence of events means dependence and independence of the corresponding points in the geometric representation. The two following criteria characterizing the linear dependence of events can now be deduced in a manner more intuitive than the direct way. In the system of barycentric coordinates, where $x_i = 1$, $x_j = 0$ $(j \neq i)$ represents the point A_i, the coordinates of the center of gravity of A_{h_1}, A_{h_2}, ..., A_{h_k} having a mass k will be

$$x_{h_1} = x_{h_2} = \cdots = x_{h_k} = 1, \qquad x_j = 0 \ (j \neq h_1, h_2, \ldots, h_k);$$

the sum of the constituents can thus be represented by a symbol of s digits, 1 or 0 (for example, the sum $C_1 \vee C_3$ by $10100 \cdots 0$). Events are linearly dependent when the matrix of the coordinates of the corresponding points and of the center 0 is of less than maximum rank, the rows of this matrix being the expressions described above corresponding to the events in question and—for the last line which consists only of 1's—the certain event. The other condition is that the events are linearly dependent when a coefficient can be assigned to each of them in such a way that in every possible case the sum of the coefficients of the events that occur always has the same value. If, in fact, the points corresponding to the given events and the point 0 are linearly dependent, it is possible to express 0 by a linear combination of the others, and this means that there exists a combination of bets on these events equivalent to a bet on the certain event.

An assignment of probability can be represented not only by the hyperplane $f = 0$ but also by a point not exterior to the simplex, conjugate to the hyperplane,[6] and defined as the center of gravity of s points having masses proportional to the probabilities of the events (constituents) that they represent. This representation is useful because the simplex gives an intuitive image

(6) In the polarity $f\binom{x}{y} = \Sigma x_i y_i = 0$ (barycentric coordinates). It is convenient here, having to employ metric notions, to consider the simplex to be equilateral. It can be specified, then, that it is a question of the polarity relative to the imaginary hypersphere $\Sigma x_i^2 = 0$, and that it makes correspond to any point A whatever the hyperplane orthogonal to the line AO passing through the point A' corresponding to A in an inversion about the center O. In vectorial notation, the hyperplane is the locus of all points Q such that the scalar product $(A - O) \cdot (Q - O)$ gives $- R^2$, where $R = l/\sqrt{2s}$, l being the length of each edge of the simplex.

of the space of probability laws, and above all because linear relations are conserved. The ∞^{s-1} admissible assignments of probability can in fact be combined linearly: if P_1, P_2, . . . , P_m are probability functions, $P = \Sigma \lambda_i P_i$ $\lambda_i \geqslant 0$, $\Sigma \lambda_i = 1$ is also, and the point representing P is given by the same relation i.e., it is the center gravity of the representative points of P_1, . . . , P_m with masses λ_1, . . . , λ_m; the admissible assignments of probability constitute then, as do the non-exterior points of the simplex, a closed, convex set. This simple remark allows us to complete our results quickly, by specifying the lack of determination of the probability of an event which remains when the event is linearly independent of certain others after the probability of the others has been fixed. It suffices to note that by fixing the value of the probability of certain events, one imposes linear conditions on the function P; the functions P that are still admissible also constitute a closed, convex set. From this one arrives immediately at the important conclusion that when the probability of an event E is not uniquely determined by those probabilities given, the admissible numbers are all those numbers in a closed interval $p' \leqslant p \leqslant p''$. If E' and E'' are respectively the sum of all the constituents contained in E or compatible with E, p' will be the smallest value admissible for the probability of E' and p'' the greatest for E''.

When the events considered are infinite in number, our definition introduces no new difficulty: P is a probability function for the infinite class of events $\&$ when it is a probability function for all finite subclasses of $\&$. This conclusion implies that the theorem of total probability cannot be extended to the case of an infinite or even denumerable number of events[7]; a discussion of this subject would carry us too far afield.

We have yet to consider the definition of conditional probabilities and the demonstration of the multiplication theorem for probabilities. Let there be two events E' and E''; we can bet on E' and condition this bet on E'': if E'' does not occur, the bet will be annulled; if E'' does occur, it will be won or lost according to whether E' does or does not occur. One can consider, then, the "conditional events" (or "tri-events"), which are the events of a three-valued logic: this "tri-event," "E' conditioned on E''," $E'|E''$, is the logical entity capable of having three values: *true* if E'' and E' are true: *false* if E'' is true and E' false; *void* if E'' is false. It is clear that two tri-events $E_1'|E_1''$ and $E_2'|E_2''$ are equal if $E_1'' = E_2''$ and $E_1'E_1'' = E_2'E_2''$; we will say that $E'|E''$ is written in *normal* form if $E' \rightarrow E''$, and it may be seen that any tri-event can be written in a single way in normal form: $E'E''|E''$. We could establish for the tri-events a three-valued logic perfectly analogous to ordinary logic [64], but this is not necessary for the goal we are pursuing.

(7)   See [16], [24], [X], [28], [XI], [64].

Let us define the probability p of E' conditioned on E'' by the same condition relative to bets, but in this case we make the convention that the bet is to be called off if E'' does not happen. The bet can then give three different results: if S is the stake, outlay paid will be pS, and the gain $(1 - p)S, - pS$, or 0 according to whether $E'|E''$ will be true, false, or void, for in the first case one gains the stake and loses the outlay, in the second one loses the outlay, and in the last the outlay is returned (if $S < 0$ these considerations remain unchanged; we need only to change the terminology of debit and credit). Let us suppose that $E' \rightarrow E''$, and let p' and p'' be the probabilities of E' and E'': we will show that for coherence we must have $p' = p \cdot p''$. If we make three bets: one on E' with the stake S', one on E'' with the stake S'', and one on $E'|E''$ with the stake S, the total gain corresponds, in the three possible cases, to

E':
$$G_1 = (1 - p') \cdot S' + (1 - p'') \cdot S'' + (1 - p)S;$$

E'' and not E':
$$G_2 = -p'S' + (1 - p'')S'' - pS;$$

not E'':
$$G_3 = -p'S' - p''S''.$$

If the determinant

$$\begin{vmatrix} 1 - p' & 1 - p'' & 1 - p \\ - p' & 1 - p'' & - p \\ - p' & - p'' & 0 \end{vmatrix} = p' - pp''$$

is not zero, one can fix S, S', and S'' in such a way that the G's have arbitrary values, in particular, all positive, and that implies a lack of coherence. Therefore $p' = pp''$, and, in general, if E' does not imply E'', this will still be true if we consider $E'E''$ rather than E': we thus have the multiplication theorem for probabilities[8]

$$P(E' \cdot E'') = P(E') \cdot P(E''|E'). \tag{1}$$

(8) This result, which, in the scheme of bets, can be deduced as we have seen from the definition of coherence, may also be expressed in a purely qualitative form, such as the following, which may be added as a fifth axiom to the preceding four (see p. 60-61): If E' and E'' are contained in E, $E'|E$ is more or less probable than (or is equal in probability to) $E''|E$, according to whether E' is more or less probable than (or equal in probability to) E''.

The condition is not only necessary, but also sufficient, in the same sense as in the case of the theorem of total probability. According to whether an individual evaluates $P(E'|E'')$ as greater than, smaller than, or equal to $P(E')$, we will say that he judges the two events to be in a positive or negative correlation, or as independent: it follows that the notion of independence or dependence of two events has itself only a subjective meaning, relative to the particular function P which represents the opinion of a given individual.

We will say that $E_1, E_2, \ldots, E_n$ constitute a class of independent events if each of them is independent of any product whatever of several others of these events (pairwise independence, naturally, does not suffice); in this case the probability of a logical product is the product of the probabilities, and, the constituents themselves being logical products, the probability of any event whatever logically dependent on $E_1, \ldots, E_n$ will be given by an algebraic function of $p_1, p_2, \ldots, p_n$.

We obtain as an immediate corollary of (1), Bayes's theorem, in the form[9]

$$P(E''|E') = \frac{P(E'') \cdot P(E'|E'')}{P(E')}, \qquad (2)$$

which can be formulated in the following particularly meaningful way: The probability of E', relative to E'', is modified in the same sense and in the same measure as the probability of E'' relative to E'.

In what precedes I have only summarized in a quick and incomplete way some ideas and some results with the object of clarifying what ought to be understood, from the subjectivistic point of view, by "logical laws of probability" and the way in which they can be proved. These laws are the conditions which characterize coherent opinions (that is, opinions admissible in their own right) and which distinguish them from others that are intrinsically contradictory. The choice of one of these admissible opinions from among all the others is not objective at all and does not enter into the logic of the probable; we shall concern ourselves with this problem in the following chapters.

(9)   It is also found expressed in this form in Kolmogorov [XVII].

CHAPTER II

The Evaluation of a Probability

The notion of probability which we have described is without doubt the closest to that of "the man in the street"; better yet, it is that which he applies every day in practical judgments. Why should science repudiate it? What more adequate meaning could be discovered for the notion?

It could be maintained, from the very outset, that in its usual sense probability cannot be the object of a mathematical theory. However, we have seen that the rules of the calculus of probability, conceived as conditions necessary to ensure coherence among the assignments of probability of a given individual, can, on the contrary, be developed and demonstrated rigorously. They constitute, in fact, only the precise expression of the rules of the logic of the probable which are applied in an unconscious manner, qualitatively if not numerically, by all men in all the circumstances of life.[e]

It can still be doubted whether this conception, which leaves each individual free to evaluate probabilities as he sees fit, provided only that the condition of coherence be satisfied, suffices to account for the more or less strict agreement which is observed among the judgments of diverse individuals, as well as between predictions and observed results. Is there, then, among the infinity of evaluations that are perfectly admissible in themselves, one particular evaluation which we can qualify, in a sense as yet unknown, as *objectively correct*? Or, at least, can we ask if a given evaluation is better than another?

There are two procedures that have been thought to provide an objective meaning for probability: the scheme of equally probable cases, and the consideration of frequencies. Indeed it is on these two procedures that the

(e)   Such a statement is misleading if, as unfortunately has sometimes happened, it is taken too seriously. It cannot be said that people compute according to arithmetic or think according to logic, unless it is understood that mistakes in arithmetic or in logic are very natural for all of us. It is still more natural that mistakes are common in the more complex realm of probability; nevertheless it seems correct to say that, fundamentally, people behave according to the rules of coherence even though they frequently violate them (just as it may be said that they accept arithmetic and logic). But in order to avoid frequent misunderstandings it is essential to point out that probability theory is not an attempt to describe actual behavior; its subject is coherent behavior, and the fact that people are only more or less coherent is inessential.

evaluation of probability generally rests in the cases where normally the opinions of most individuals coincide. However, these same procedures do not oblige us at all to admit the existence of an objective probability; on the contrary, if one wants to stretch their significance to arrive at such a conclusion, one encounters well-known difficulties, which disappear when one becomes a little less demanding, that is to say, when one seeks not to eliminate but to make more precise the subjective element in all this. In other words, it is a question of considering the coinicidence of opinions as a psychological fact; the reasons for this fact can then retain their subjective nature, which cannot be left aside without raising a host of questions of which even the sense is not clear. Thus in the case of games of chance, in which the calculus of probability originated, there is no difficulty in understanding or finding very natural the fact that people are generally agreed in assigning equal probabilities to the various possible cases, through more or less precise, but doubtless very spontaneous, considerations of symmetry. Thus the classical definition of probability, based on the relation of the number of favorable cases to the number of possible cases, can be justified immediately: indeed, if there is a complete class of n incompatible events, and if they are judged equally probable, then by virtue of the theorem of total probability each of them will necessarily have the probability $p = 1/n$ and the sum of m of them the probability m/n. A powerful and convenient criterion is thus obtained: not only because it gives us a way of calculating the probability easily when a subdivision into cases that are judged equally probable is found, but also because it furnishes a general method for evaluating by comparison any probability whatever, by basing the quantitative evaluation on purely qualitative judgments (equality or inequality of two probabilities). However this criterion is only applicable on the hypothesis that the individual who evaluates the probabilities judges the cases considered equally probable; this is again due to a subjective judgment for which the habitual considerations of symmetry which we have recalled can furnish psychological reasons, but which cannot be transformed by them into anything objective. If, for example, one wants to demonstrate that the evaluation in which all the probabilities are judged equal is alone "right," and that if an individual does not begin from it he is "mistaken," one ought to begin by explaining what is meant by saying that an individual who evaluates a probability judges "right" or that he is "mistaken." Then one must show that the conditions of symmetry cited imply necessarily that one must accept the hypothesis of equal probability if one does not want to be "mistaken." But any event whatever can only happen or not happen, and neither in one case nor in the other can one decide what would be the degree of doubt with which it would be "reasonable" or "right" to expect the event before knowing whether it has occurred or not.

Let us now consider the other criterion, that of frequencies. Here the problem is to explain its value from the subjectivistic point of view and to show precisely how its content is preserved. Like the preceding criterion, and like all possible criteria, it is incapable of leading us outside the field of subjective judgments; it can offer us only a more extended psychological analysis. In the case of frequencies this analysis is divided into two parts: an elementary part comprised of the relations between evaluations of probabilities and predictions of future frequencies, and a second, more delicate part concerning the relation between the observation of past frequencies and the prediction of future frequencies. For the moment we will limit ourselves to the first question, while admitting as a known psychological fact, whose reasons will be analyzed later, that one generally predicts frequencies close to those that have been observed.

The relation we are looking for between the evaluation of probabilities and the prediction of frequencies is given by the following theorem. Let E_1, $E_2, \ldots, E_n$ be any events whatever.[1] Let us assign the values $p_1, p_2, \ldots, p_n$ to their probabilities and the values $\omega_0, \omega_1, \ldots, \omega_n$, to the probabilities that zero, or only one, or two, etc., or finally, all these events will occur (clearly $\omega_0 + \omega_1 + \omega_2 + \cdots + \omega_n = 1$). For coherence, we must have:

$$p_1 + p_2 + \cdots + p_n = 0 \times \omega_0 + 1 \times \omega_1 + 2 \times \omega_2 + \cdots + n \times \omega_n$$

or simply

$$\bar{p} = \bar{f} \qquad (3)$$

where $\bar{p}$ indicates the arithmetic mean of the p_i, and $\bar{f}$ the mathematical

(1)   In order to avoid a possible misunderstanding due to the divergence of our conception from some commonly accepted ones, it will be useful to recall that, in our terminology, an "event" is always a determinate singular fact. What are sometimes called *repetitions* or *trials* of the same event are for us distinct events. They have, in general, some common characteristics or symmetries which make it natural to attribute to them equal probabilities, but we do not admit any *a priori* reason which prevents us in principle from attributing to each of these trials $E_1, \ldots,$ E_n some different and absolutely arbitrary probabilities $p_1, \ldots, p_n$. In principle there is no difference for us between this case and the case of n events which are not analogous to each other; the analogy which suggests the name "trials of the same event" (we would say "of the same phenomenon") is not at all essential, but, at the most, valuable because of the influence it can exert on our psychological judgment in the sense of making us attribute equal or very nearly equal probabilities to the different events.

expectation of the frequency (that is to say of the random quantity which takes the values $0/n$, $1/n$, $2/n$, ..., n/n according to whether $0, 1, 2, ..., n$ of the E_i occur); we note that in this respect the notion of mathematical expectation has itself a subjective meaning, since it is defined only in relation to the given judgment which assigns to the $n + 1$ possible cases the probabilities ω_h.

This relation can be further simplified in some particular cases: if the frequency is known, the second member simply represents that value of the frequency; if one judges that the n events are equally probable, the first member is nothing but the common value of the probability. Let us begin with the case in which both simplifying assumptions are correct: there are n events, m are known to have occurred or to be going to occur, but we are ignorant of which, and it is judged equally probable that any one of the events should occur. The only possible evaluation of the probability in this case leads to the value $p = m/n$. If $m = 1$, this reduces to the case of n equally probable, incompatible possibilities.

If, in the case where the frequency is known in advance, our judgment is not so simple, the relation is still very useful to us for evaluating the n probabilities, for by knowing what their arithmetic mean has to be, we have a gross indication of their general order of magnitude, and we need only arrange to augment certain terms and diminish others until the relation between the various probabilities corresponds to our subjective judgment or the inequality of their respective chances. As a typical example, consider a secret ballot: one knows that among the n voters A_1, A_2, ..., A_n, one has m favorable ballots; one can then evaluate the probabilities p_1, p_2, ..., p_n that the different voters have given a favorable vote, according to the idea one has of their opinions; in any case this evaluation must be made in such a way that the arithmetic mean of the p_i will be m/n.

When the frequency is not known, the equation relates two terms which both depend on a judgment of probability: the evaluation of the probabilities p_i is no longer bound by their average to something given objectively, but to the evaluation of other probabilities, the probabilities ω_h of the various frequencies. Still, it is an advantage not to have to evaluate exactly all the ω_h in order to apply the given relation to the evaluation of the probabilities p_i; a very vague estimation of a qualitative nature suffices, in fact, to evaluate $\bar{f}$ with enough precision. It suffices, for example, to judge as "not very probable" that the frequency differs noticeably from a certain value a, which is tantamount to estimating as very small the sum of all the ω_h for which $|h/n - a|$ is not small, to give approximately $\bar{f} = a$.

Once $\bar{f}$ has been evaluated, nothing is changed of what we said earlier concerning the case where the frequency is known: if the n events are judged equally probable, their common probability is $p = \bar{f}$; if that is not the case, then certain probabilities will be augmented or diminished in order that their arithmetic mean will be $\bar{f}$.

It is thus that one readily evaluates probabilities in most practical problems, for example, the probability that a given individual, let us say Mr. A, will die in the course of the year. If it is desired to estimate directly under these conditions what stakes (or insurance, as one would prefer to say in this case) seem to be equitable, this evaluation would seem to us to be affected with great uncertainty; the application of the criterion described above facilitates the estimation greatly. For this one must consider other events, for example, the deaths, during the year, of individuals of the same age and living in the same country as Mr. A. Let us suppose that among these individuals about 13 out of 1000 will die in a year; if, in particular, all the probabilities are judged equal, their common value is $p = 0.013$, and the probability of death for Mr. A is 0.013; if in general there are reasons which make the chances we attribute to their deaths vary from one individual to another, this average value of 0.013 at least gives us a base from which we can deviate in one direction or the other in taking account of the characteristics which differentiate Mr. A from other individuals.

This procedure has three distinct and successive phases: the first consists of the choice of a class of events including that which we want to consider; the second is the prediction of the frequency; the third is the comparison between the average probability of the single events and that of the event in question. Some observations in this regard are necessary in order to clarify the significance and value that are attributed to these considerations by subjectivists' point of view, and to indicate how these views differ from current opinion. Indeed, it is only the necessity of providing some clarification about these points before continuing that makes it indispensable to spend some little time on such an elementary question.

The choice of a class of events is in itself arbitrary; if one chooses "similar" events, it is only to make the application of the procedure easier, that is to say, to make the prediction of the frequency and the comparison of the various probabilities easier: but this restriction is not at all essential, and even if one admits it, its meaning is still very vague. In the preceding example, one could consider, not individuals of the same age and the same country, but those of the same profession, of the same height, of the same profession and town, etc., and in all these cases one could observe a noticeable enough similarity. Nothing prevents *a priori* the grouping of the event which interests us with any other events whatever. One can consider, for example, the death of Mr. A during the year as a *claim* in relation to all the policies of the company by which he is insured, comprising fire insurance, transport insurance, and others; from a certain point of view, one can still maintain that these events are "similar."

This is why we avoid expressions like "trials of the same event," "events which can be repeated," etc., and, in general, all the frequency considerations which presuppose a classification of events, conceived as rigid and essential, into classes or collections or series. All classifications of this sort have only an auxiliary function and an arbitrary value.

The prediction of the frequency is based generally on the hypothesis that its value remains nearly constant: in our example, the conviction that the proportion of deaths is 13 per 1000 can have its origin in the observation that in the course of some years past the mortality of individuals of the same kind was in the neighborhood of 13/1000. The reasons which justify this way of predicting could be analyzed further; for the moment it suffices to assume that in effect our intuition leads us to judge thus. Let us remark that such a prediction is generally the more difficult the narrower the class considered.

On the other hand, the comparison of the different probabilities is more difficult in the same proportion the events are more numerous and less homogeneous: the difficulty is clearly reduced to a minimum when the events appear to us equally probable. In practice one must attempt to reconcile as best one can these opposing demands, in order to achieve the best application of the two parts of the procedure: it is only as a function of these demands that the class of events considered can be chosen in a more or less appropriate fashion.

An illustration will render these considerations still clearer. If one must give an estimate of the thickness of a sheet of paper, he can very easily arrive at it by estimating first the thickness of a packet of n sheets in which it is inserted, and then by estimating the degree to which the various sheets have the same thickness. The thickness can be evaluated the more easily the larger the packet; the difficulty of the subsequent comparison of the sheets is on the contrary diminished if one makes the packet thinner by saving only those sheets judged to have about the same thickness as the sheet that interests us.

Thus the criterion based on the notion of frequency is reduced, like that based on equiprobable events, to a practical method for linking certain subjective evaluations of probability to other evaluations, themselves subjective, but preferable either because more accessible to direct estimation, or because a rougher estimate or even one of a purely qualitative nature suffices for the expected conclusions. *A priori*, when one accepts the subjectivistic point of view, such ought to be the effective meaning and the value of any criterion at all.

In the case of predictions of frequencies, one only relates the evaluation of p_i to that of the ω_h and to a comparison between the p_i; the estimation of the ω_h does not need to come up to more than a rough approximation, such as suffices to determine the p_i closely enough. It must be remarked nevertheless that this prediction of the frequency is nothing else than an evaluation of the ω_h; it is not a prophecy which one can call correct if the frequency is equal or close to $\bar{f}$, and false in the contrary case. All the frequencies $0/n$, $1/n$, $2/n$, . . . , n/n are possible, and whatever the realized frequency may be, nothing can make us right or wrong if our actual judgment is to attribute to these $n + 1$ cases the probabilities ω_h, leading to a certain value

$$\bar{p} = \bar{f} = \frac{\omega_1 + 2\omega_2 + 3\omega_3 + \cdots + n\omega_n}{n}. \tag{3}$$

It is often thought that these objections may be escaped by observing that the impossibility of making the relations between probabilities and frequencies precise is analogous to the practical impossibility that is encountered in all the experimental sciences of relating exactly the abstract notions of the theory and the empirical realities.[2] The analogy is, in my view, illusory: in the other sciences one has a theory which asserts and predicts with certainty and exactitude what would happen if the theory were completely exact; in the calculus of probability it is the theory itself which obliges us to admit the possibility of all frequencies. In the other sciences the uncertainty flows indeed from the imperfect connection between the theory and the facts; in our case, on the contrary, it does not have its origin in this link, but in the body of the theory itself [32], [65], [IX]. No relation between probabilities and frequencies has an empirical character, for the observed frequency, whatever it may be, is always compatible with all the opinions concerning the respective probabilities; these opinions, in consequence, can be neither confirmed nor refuted, once it is admitted that they contain no categorical assertion such as: such and such an event *must* occur or *can not* occur.

This last consideration may seem rather strange if one reflects that the prediction of a future frequency is generally based on the observation of those past; one says, "we will correct" our initial opinions if "experience refutes them." Then isn't this instinctive and natural procedure justified? Yes; but the way in which it is formulated is not exact, or more precisely, is not meaningful. It is not a question of "correcting" some opinions which have been "refuted"; it is simply a question of substituting for the initial evaluation of the probability the value of the probability which is conditioned on the occurrence of facts which have already been observed; this probability is a completely different thing from the other, and their values

(2) This point of view is maintained with more or less important variations in most modern treatises, among others those of Castelnuovo [VI], Fréchet-Halbwachs [XII], Lévy [XX], von Mises [XXV].

can very well not coincide without this non-coincidence having to be interpreted as the "correction of a refuted opinion."

The explanation of the influence exercised by experience on our future
predictions, developed according to the ideas that I have just expounded,
constitutes the point that we have left aside in the analysis of the criterion
based on frequencies. This development will be the subject of the following
chapters, in which we will make a more detailed study of the most typical
case in this connection: the case of exchangeable events, and, in general, of
any exchangeable random quantities or elements whatever. This study is
important for the development of the subjectivistic conception, but I hope
that the mathematical aspect will be of some interest in itself, independently
of the philosophical interpretation; in fact, exchangeable random quantities
and exchangeable events are characterized by simple and significant conditions which can justify by themselves a deep study of the problems that arise
in connection with them.

CHAPTER III

Exchangeable Events

Why are we obliged in the majority of problems to evaluate a probability
according to the observation of a frequency? This is a question of the relations between the observation of past frequencies and the prediction of
future frequencies which we have left hanging, but which presents itself anew
under a somewhat modified form when we ask ourselves if a prediction of
frequency can be in a certain sense confirmed or refuted by experience. The
question we pose ourselves now includes in reality the problem of reasoning
by induction. Can this essential problem, which has never received a satisfactory solution up to now, receive one if we employ the conception of subjective probability and the theory which we have sketched?

In order to fix our ideas better, let us imagine a concrete example, or
rather a concrete interpretation of the problem, which does not restrict its
generality at all. Let us suppose that the game of heads or tails is played with
a coin of irregular appearance. The probabilities of obtaining "heads" on the
first, the second, the hth toss, that is to say, the probabilities $P(E_1)$, $P(E_2)$,
..., $P(E_h)$, ... of the events $E_1, E_2, \ldots, E_h, \ldots$ consisting of the occurrence of heads on the different tosses, can only be evaluated by calculating
a priori the effect of the apparent irregularity of the coin.

It will be objected, no doubt, that in order to get to this point, that is to say, to obtain the "correct" probabilities of future trials, we can utilize the results obtained in the previous trials: it is indeed in this sense that—according to the current interpretation—we "correct" the evaluation of $P(E_{n+1})$ after the observation of the trials which have, or have not, brought about $E_1, E_2, \ldots, E_n$. Such an interpretation seems to us unacceptable, not only because it presupposes the objective existence of unknown probabilities, but also because it cannot even be formulated correctly: indeed the probability of E_{n+1} evaluated with the knowledge of a certain result, A, of the n preceding trials is no longer $P(E_{n+1})$ but $P(E_{n+1} | A)$. To be exact, we will have

$$A = E_{i_1} E_{i_2} \cdots E_{i_r} \overline{E}_{j_1} \overline{E}_{j_2} \cdots \overline{E}_{j_s} \ (r + s = n),$$

the result A consisting of the r throws $i_1, i_2, \ldots, i_r$ giving "heads" and the other s throws $j_1, j_2, \ldots, j_s$ giving tails: A is then one of the constituents formed with $E_1, E_2, \ldots, E_n$. But then, if it is a question of a conditional probability, we can apply the theorem of compound probability, and the interpretation of the results which flow from this will constitute our justification of inductive reasoning.

In general, we have

$$P(E_{n+1} | A) = \frac{P(A \cdot E_{n+1})}{P(A)} ; \tag{4}$$

our explanation of inductive reasoning is nothing else, at bottom, than the knowledge of what this formula expresses: the probability of E_{n+1} evaluated when the result A of $E_1, \ldots, E_n$ is known, is not something of an essentially novel nature (justifying the introduction of a new term like "statistical" or "*a posteriori*" probability). This probability is not independent of the "*a priori* probability" and does not replace it; it flows in fact from the same *a priori* judgment by subtracting, so to speak, the components of doubt associated with the trials whose results have been obtained.[f]

(f)   This terminology derives from the time when a philosophical distinction was made between probabilities evaluated by considerations of symmetry (*a priori* probabilities), and those justified statistically (*a posteriori* probabilities); this dualistic view is now rejected not only in the subjectivistic theory maintained here, but also by most authors of other theories. With reference to current views, it is proper to speak simply of *initial* and *final* probabilities (the difference being relative to a particular problem where one has to deal with evaluations at different times, before and after some specific additional information has been obtained); the terminology has not been modernized here because the passage makes reference to the older views.

In order to avoid erroneous interpretations of what follows, it is best at the outset to recall once more the sense which we attribute to a certain number of terms in this work. Let us consider, to begin with, a class of events (as, for example, the various tosses of a coin). We will say sometimes that they constitute the *trials* of a given phenomenon; this will serve to remind us that we are almost always interested in applying the reasoning that follows to the case where the events considered are events *of the same type,* or which have *analogous* characteristics, without attaching an intrinsic significance or a precise value to these exterior characteristics whose definition is largely arbitrary. Our reasoning will only bring in the events, that is to say, the trials, each taken individually; the analogy of the events does not enter into the chain of reasoning in its own right but only to the degree and in the sense that it can influence in some way the judgment of an individual on the probabilities in question.

It is evident that by posing the problem as we have, it will be impossible for us to *demonstrate* the validity of the principle of induction, that is to say, the principle according to which the probability ought to be close to the observed frequency—for example, in the preceding case: $P(E_{n+1} \mid A) \cong r/n$. That this principle can only be justified in particular cases is not due to an insufficiency of the method followed, but corresponds logically and necessarily to the essential demands of our point of view. Indeed, probability being purely subjective, nothing obliges us to choose it close to the frequency; all that can be shown is that such an evaluation follows in a coherent manner from our initial judgment when the latter satisfies certain perfectly clear and natural conditions.

We will limit ourselves in what follows to the simplest conditions which define the events which we call exchangeable, and to fix our ideas we will exhibit these conditions in the example already mentioned; our results will nevertheless be completely general.

The problem is to evaluate the probabilities of all the possible results of the n first trials (for any n). These possible results are 2^n in number, of which

$\binom{n}{n} = 1$ consist of the repetition of "heads" n times, $\binom{n}{n-1} = n$ of $n - 1$ occurrences of "heads" and one occurrence of "tails," ..., and in general

$\binom{n}{r}$ of r occurrences of "heads" and $n - r$ occurrences of "tails." If we designate by $\omega_r^{(n)}$ the probability that one obtains in n tosses, in any order whatever, r occurrences of "heads" and $n - r$ occurrences of "tails," $\omega_r^{(n)}$ will be the sum of the probabilities of the $\binom{n}{r}$ distinct ways in which one can obtain this result; the average of these probabilities will then be $\omega_r^{(n)} \Big/ \binom{n}{r}$.

Having grouped the 2^n results in this way, we can distinguish usefully, though arbitrarily, two kinds of variation in the probabilities: to begin with we have an average probability which is greater or smaller for each frequency, and then we have a more or less uniform subdivision of the probabilities $\omega_r^{(n)}$ among the various results of equal frequency that only differ from one another in the order of succession of favorable and unfavorable trials. In general, different probabilities will be assigned, depending on the order, whether it is supposed that one toss has an influence on the one which follows it immediately, or whether the exterior circumstances are supposed to vary, etc.; nevertheless it is particularly interesting to study the case where the probability does not depend on the order of the trials. In this case every result having the same frequency r/n on n trials has the same probability, which is $\omega_r^{(n)} \Big/ \binom{n}{r}$; if this condition is satisfied, we will say that the events of the class being considered, e.g., the different tosses in the example of tossing coins, are *exchangeable* (in relation to our judgment of probability). We will see better how simple this condition is and the extent to which its significance is natural, when we have expressed it in other forms, some of which will at first seem more general, and others more restrictive.

It is almost obvious that the definition of exchangeability leads to the following result: the probability that n determinate trials will all have a favorable result is always the same, whatever the n-tuple chosen: this probability will be equal to $\omega_n = \omega_n^{(n)}$, since the first n cases constitute a particular n-tuple. Conversely, if the probabilities of the events have this property, the events are exchangeable, for, as will be shown a little later, it follows from this property that all the results having r favorable and s unfavorable results out of n trials have the same probability, that is:

$$\frac{\omega_r^{(n)}}{\binom{n}{r}} = (-1)^s \Delta^s \omega_r \qquad (5)$$

Another conclusion has already been obtained: the probability that r trials will be favorable and s unfavorable will always be $\omega_r^{(n)} \Big/ \binom{n}{r}$ (with $n = r + s$), not only when it is a question of the first n trials in the original order, but also in the case of any trials whatever.

Another condition, equivalent to the original definition, can be stated: the probability of any trial E whatever, conditional on the hypothesis A that there have been r favorable and s unfavorable results on the other specific trials, does not depend on the events chosen, but simply on r and s (or on

r and $n = r + s$).[g] If

$$P(A) = \frac{\omega_r^{(n)}}{\binom{n}{r}} \quad \text{and} \quad P(A \cdot E) = \frac{\omega_{r+1}^{(n+1)}}{\binom{n+1}{r+1}}$$

then we will have

$$P(E \mid A) = \frac{r+1}{n+1} \left(\frac{\omega_{r+1}^{(n+1)}}{\omega_r^{(n)}} \right) = p_r^{(n)} \tag{6}$$

a function of n and r only; if, on the other hand, one supposes that $P(E \mid A) = p_r^{(n)}$, a function of n and r only, it follows clearly that for every n-tuple the probability that all the trials will be favorable is

$$\omega_n = p_0^{(0)} \cdot p_1^{(1)} \cdots p_{n-1}^{(n-1)}. \tag{7}$$

In general it may easily be seen that in the case of exchangeable events, the whole problem of probabilities concerning E_{i_1}, E_{i_2}, ..., E_{i_n} does not depend on the choice of the (distinct) indices $i_1, \ldots, i_n$, but only on the probabilities $\omega_0, \omega_1, \ldots, \omega_n$. This fact justifies the name of "exchangeable events" that we have introduced: when the indicated condition is satisfied, any problem is perfectly well determined if it is stated for *generic* events. This same fact makes it very natural to extend the notion of exchangeability to the larger domain of random quantities: We shall say that $X_1, X_2, \ldots,$ $X_n, \ldots$ are exchangeable random quantities if they play a symmetrical role in relation to all problems of probability, or, in other words, if the probability that $X_{i_1}, X_{i_2}, \ldots, X_{i_n}$ satisfy a given condition is always the same however the distinct indices $i_1 \cdots i_n$ are chosen. As in the case for exchangeable *events,* any problem of probability is perfectly determined when it has been stated for *generic* random quantities; in particular if $X_1, X_2, \ldots, X_n,$ $\ldots$ are exchangeable random quantities, the events $E_i = (X_i \leqslant x)$ (where x is any fixed number) or more generally $E_i = (X_i \in I)$ (I being any set of numbers) are exchangeable. This property will be very useful to us, as in the following case: the mathematical expectation of any function of n exchangeable random quantities does not change when we change the n-tuple chosen; in particular there will be values $m_1, m_2, \ldots, m_k, \ldots$ such that $\mathcal{M}(X_i) = m_1$,

(g)   This may also be expressed by saying that the observed frequency r/n and n give a *sufficient statistic,* or that the likelihood is only a function of r/n and n.

whatever i may be; $\mathscr{M}(X_i X_j) = m_2$, whatever be i and j $(i \neq j)$, and in general $\mathscr{M}(X_{i_1} X_{i_2} \cdots X_{i_k}) = m_k$ whatever be the distinct $i_1, i_2, \ldots, i_k$. This observation has been made by Kinchin[1] who has used it to simplify the proofs of some of the results that I have established for exchangeable events. I have used this idea in the study of exchangeable random quantities, and I will avail myself of it equally in this account.

One can, indeed, treat the study of exchangeable events as a special case of the study of exchangeable random quantities, by observing that the events E_i are exchangeable only if that is also true of their "indicators," that is to say, the random quantities X_i such that $X_i = 1$ or $X_i = 0$ according to whether E_i occurs or not. We mention in connection with these "indicators" some of the simple properties which explain their usefulness.

The indicator of $\overline{E}_i$ is $1 - X_i$; that of $E_i E_j$ is $X_i X_j$; that of $E_i \vee E_j$ is $1 - (1 - X_i)(1 - X_j) = X_i + X_j - X_i X_j$—it is not, as it is in the case of incompatible events where $X_i X_j = 0$, simply $X_i + X_j$. The indicator of $E_{i_1} E_{i_2} \cdots E_{i_r} \overline{E}_{j_1} \overline{E}_{j_2} \ldots \overline{E}_{j_s}$ is then

$$X_{i_1} X_{i_2} \cdots X_{i_r} (1 - X_{j_1})(1 - X_{j_2}) \cdots (1 - X_{j_s})$$

$$= X_{i_1} X_{i_2} \cdots X_{i_r} - \sum_{h=1}^{s} X_{i_1} X_{i_2} \cdots X_{i_r} X_{j_h}$$

$$+ \sum_{k,h=1}^{s} X_{i_1} X_{i_2} \cdots X_{i_r} X_{j_h} X_{j_k} - \cdots \pm X_1 X_2 \cdots X_n$$

The mathematical expectation of the indicator is only the probability of the corresponding event; thus the possibility of transforming the logical operations on the events into arithmetical operations on the indicators greatly facilitates the solution of a certain number of problems. One infers immediately, in particular, the formula (5) stated for $\omega_r^{(n)}$ in the case of exchangeable events: if the product of h trials always has the probability ω_h, then the probability $\omega_r^{(n)} / \binom{n}{r}$ of $E_{i_1} E_{i_2} \cdots E_{i_r} \overline{E}_{j_1} \overline{E}_{j_2} \cdots \overline{E}_{j_s}$ is deduced from the above development of the indicator of this event and one obtains

$$\frac{\omega_r^{(n)}}{\binom{n}{r}} = \omega_r - \binom{s}{1} \omega_{r+1} + \binom{s}{2} \omega_{r+2} - \cdots (-1)^s \omega_{r+s} = (-1)^s \Delta^s \omega_r. \quad (5)$$

Putting $\omega_0 = 1$, the formula remains true for $r = 0$.

(1) [XV]; also see [XVI].

Leaving aside for the moment the philosophical question of the principles which have guided us here, we will now develop the study of exchangeable events and exchangeable random quantities, showing first that the law of large numbers and even the strong law of large numbers are valid for exchangeable random quantities X_i, and that the probability distribution of the average Y_n of n of the random quantities X_i tends toward a limiting distribution when n increases indefinitely. It suffices even, in the demonstration, to suppose

$$\mathcal{M}(X_i) = m_1, \qquad \mathcal{M}(X_i^2) = \mu_2, \qquad \mathcal{M}(X_i X_j) = m_2$$

for all i and j $(i \neq j)$, a condition which is much less restrictive than that of exchangeability. We remark again that it suffices to consider explicitly random quantities, the case of events being included by the consideration of "indicators"; an average Y_n is identical, in this case, with the frequency on n trials.

The "law of large numbers" consists of the following property: *if Y_h and Y_k are respectively the averages of h and of k random quantities X_i (the two averages may or may not contain some terms in common), the probability that $|Y_h - Y_k| > \epsilon$ ($\epsilon > 0$) may be made as small as we wish by taking h and k sufficiently large;* this follows immediately from the calculation of the mathematical expectation of $(Y_h - Y_k)^2$:

$$\mathcal{M}(Y_h - Y_k)^2 = \frac{h + k - 2r}{hk} \, (\mu_2 - m_2)$$

$$= \left(\frac{1}{h} + \frac{1}{k} - \frac{2r}{hk} \right) (\mu_2 - m_2) \leqslant \left(\frac{1}{h} + \frac{1}{k} \right) (\mu_2 - m_2), \quad (8)$$

where r is the number of common terms, i.e., the X_i that occur in Y_h as well as in Y_k. In particular, if it is a question of "successive" averages, that is to say, if all the terms in the first expression appear also in the other, as for example if

$$Y_h = (1/h)(X_1 + X_2 + \cdots + X_h), \, Y_k = (1/k)(X_1 + X_2 + \cdots + X_k) \, (h < k)$$

we will have $r = h$, and

$$\mathcal{M}(Y_h - Y_k)^2 = \left(\frac{1}{h} - \frac{1}{k} \right) (\mu_2 - m_2). \tag{9}$$

When successive averages are considered, we have in addition the following result, which constitutes the strong law of large numbers: *ϵ and θ being given,* *it suffices to choose h sufficiently great in order that the probability of finding the successive averages* $Y_{h+1}, Y_{h+2}, \ldots, Y_{h+q}$ *all between* $Y_h - \epsilon$ *and* $Y_h + \epsilon$ *differs from unity by a quantity smaller than* θ, *q being as great as one wants.* If one admits that the probability that all the inequalities

$$|Y_h - Y_{h+i}| < \epsilon \quad (i = 1, 2, 3, \ldots)$$

are true is equal to the limit of the analogous probability of $i = 1, 2, \ldots, q$, when $q \to \infty$, then one can say that *all* the averages Y_{h+i} $(i = 1, 2, \ldots)$ fall between $Y_h - \epsilon$ and $Y_h + \epsilon$, excepting in a case whose probability is less than θ; I prefer however to avoid this sort of statement, for it presupposes essentially the extension of the theorem of total probabilities to the case of a denumerably infinite number of events, and this extension is not admissible, at least according to my point of view (see p. 68).

The proof of the strong law of large numbers can be obtained easily, by considering the variation among the terms Y_{h+i} with the index $(h + i)$ square, and then the variation in the segments between two successive square indices. If the Y's with square indices do not differ among themselves by more than $\epsilon/3$, and the Y's with indices falling between two successive square indices do not differ from each other by more than $\epsilon/3$, the deviations among the Y_{h+i} obviously cannot exceed ϵ. But it suffices to apply the Bienaymé-Tchebycheff inequality to succeed in overestimating the probability of an exception to one of these partial limitations,[2] and the corresponding probabilities come out less than $36(\mu_2 - m_2)\epsilon^{-2} \sum\limits_{i=s}^{\infty} i^{-2}$ (s = the integral part of $\sqrt{h}$); the probability of an exception to one or the other of the partial limitations cannot therefore exceed

$$72(\mu_2 - m_2)\epsilon^{-2} \sum\limits_{i=s}^{\infty} i^{-2}.$$

This value does not depend on q, and tends toward zero when $s \to \infty$ (that is to say, when $h \to \infty$); the strong law of large numbers is thus demonstrated.

(2)   The formula $\mathcal{M}(Y_h - Y_k)^2 = \dfrac{k - h}{hk}(\mu_2 - m_2)$ gives, by the Bienaymé-Tchebycheff inequality, $P(|Y_h - Y_k| > \epsilon) < \dfrac{1}{\epsilon^2}(\mu_2 - m_2)\dfrac{k - h}{hk}$;

applying the theorem of total probabilities in the manner indicated in my note [47], it is possible to draw from this inequality the conclusions that follow in the text.

From the fact that the law of large numbers holds, the other result stated follows easily: *the distribution* $\Phi_n(\xi) = P(Y_n \leqslant \xi)$ *tends to a limit as* $n \to \infty$. If the probability that $|Y_h - Y_k| > \epsilon$ is smaller than θ, the probability that $Y_h \leqslant \xi$ and $Y_k > \xi + \epsilon$ will *a fortiori* be smaller than θ, and one will have $\Phi_h(\xi) < \Phi_k(\xi + \epsilon) + \theta$, and similarly $\Phi_h(\xi) > \Phi_k(\xi - \epsilon) - \theta$. As ϵ and θ can be chosen as small as we wish, it follows that there exists a limiting distribution $\Phi(\xi)$ such that $\lim_{n\to\infty} \Phi_n(\xi) = \Phi(\xi)$ except perhaps for points of discontinuity.[3]

If, in particular, the random quantities X_1, X_2, ..., X_n, ... are the indicators of exchangeable trials of a given phenomenon, that is to say, if they correspond to the exchangeable events E_1, E_2, ..., E_n, ..., the hypothesis will be satisfied; it would suffice even that the events be equally probable $[P(E_i) = \mathcal{M}(X_i) = m_1 = \omega_1]$ and have the same two-by-two correlation $[P(E_i E_j) = \mathcal{M}(X_i X_j) = m_2 = \omega_2]$. We remark that for the indicators one has $X^2 = X$ (since $0^2 = 0$ and $1^2 = 1$) so that $\mu_2 = m_1 = \omega_1$. For Y_h, the frequency on h trials, we then have

$$\mathcal{M}(Y_h) = \omega_1, \quad \mathcal{M}(Y_h - Y_k)^2 = (1/h + 1/k - 2r/hk)(\omega_1 - \omega_2) \qquad (10)$$

and the demonstrated results show simply that the frequencies of two sufficiently numerous groups of trials are, almost surely, very close [even if it is a question of disjoint groups $(r = 0)$; if there are some common events $(r > 0)$, so much the better]. The same results further signify that the successive frequencies in the same sequence of experiments oscillate almost surely with a quantity less than a given ϵ, beginning from a rank h sufficiently large, whatever be the number of subsequent events; and finally that there exists a probability distribution $\Phi(\xi)$ differing only slightly from that of a frequency Y_h for very large h.

In order to determine completely the limiting distribution $\Phi(\xi)$, the knowledge of m_1, m_2, μ_2, is evidently no longer sufficient, except in the limiting case where there is no two-by-two correlation and $m_2 = m_1{}^2$; here $\Phi(\xi)$ is degenerate and reduces to the distribution where the probability is concentrated in one point $\xi = m_1$. In this case the law of large numbers

(3)   We remark that if the X_i are exchangeable, the distribution $\Phi_n(\xi)$ is the same for all the averages Y_n of n terms; one then has a sequence of functions Φ_n depending solely on n and tending toward Φ; with a less restrictive hypothesis than the demonstration assumes, two averages Y_n' and Y_n'' formed from distinct terms can have two different distributions Φ_n' and Φ_n'', but the result will still hold in the sense that all the $\Phi_n(\xi)$ concerning the average of any n terms whatever will differ very little from $\Phi(\xi)$ (and thus from one another) if n is sufficiently large.

and the strong law of large numbers reduce to the laws of Bernoulli and Cantelli [III], [V], according to which the deviation between Y_h and the value m_1, fixed in advance, tends stochastically toward zero in a "strong" way. In the general case of a class of exchangeable random quantities, Φ is determined by the knowledge of the complete sequence $m_1, m_2, \ldots, m_n,$ $\ldots$, for these values are the *moments* relative to the distribution Φ:

$$m_n = \int_0^1 \xi^n d\Phi(\xi) \tag{11}$$

and then

$$\psi(t) = \sum_{0=n}^{\infty} \frac{i^n t^n}{n!} m_n \tag{12}$$

is the characteristic function of Φ.

Indeed,

$$Y_h{}^n = \frac{1}{h^n} (X_1 + X_2 + \cdots + X_n)^n = \frac{1}{h^n} \sum X_{i_1} X_{i_2} \cdots X_{i_n};$$

among the h^n products there are $h(h-1)(h-2) \cdots (h-n+1)$ that are formed from distinct factors; the products containing the same term more than one time constitute a more and more negligible fraction as h is increased, so that

$$\mathcal{M}(Y_h{}^n) = \frac{h(h-1) \cdots (h-n+1)}{h^n} m_n + \mathcal{O}\left(\frac{1}{h}\right) \to m_n \quad (h \to \infty). \tag{13}$$

If, in particular, the X_i are the indicators of exchangeable trials of a phenomenon, Y_h the frequency on h trials, and m_n is the probability ω_n that n trials will all have a favorable result, then (13) evaluates the mean of the nth power of the relative frequency on a large number of trials. The characteristic function of $\Phi(\xi)$ is

$$\psi(t) = \sum_{n=0}^{\infty} \frac{i^n t^n}{n!} \omega_n \tag{14}$$

and then we have

$$\Phi(\xi) = \frac{1}{2\pi} \int_{-\infty}^{\infty} \frac{e^{it} - e^{-it\xi}}{it} \psi(t) \, dt \tag{15}$$

for—the Y_h signifying frequencies—the probability distribution can only fall between 0 and 1, and thus $\Phi(-1) = 0$. The characteristic function of $\Phi_h(\xi)$ is

$$\psi_h(t) = \Omega_h\left(e^{i\frac{t}{h}}\right), \tag{16}$$

where Ω_h is the polynomial

$$\Omega_h(z) = \sum_{k=0}^{h} \binom{h}{k} \omega_k (z-1)^k, \tag{17}$$

and $\Omega_h(t)$ converges uniformly to $\psi(t)$. This fact can be proved directly; it is from this standpoint that I developed systematically the study of exchangeable events in my first works [29], [40], and demonstrated the existence of the limiting distribution Φ, and of ψ, which I call the "characteristic function of the phenomenon."[4]

To give the limiting distribution Φ, or the characteristic function ψ, is, as we have seen, equivalent to giving the sequence ω_n; it follows that this suffices to determine the probability for any problem definable in terms of exchangeable events. All such problems lead, indeed, in the case of exchangeable events, to the probabilities $\omega_r^{(n)}$ that on n trials, a number r will be favorable; we have (putting $s = n - r$)

$$\omega_r^{(n)} = (-1)^s \binom{n}{r} \Delta^s \omega_r = \binom{n}{r} \int_0^1 \xi^r (1-\xi)^s \, d\Phi(\xi), \tag{18}$$

and an analogous formula having the same significance is valid for the general case. Indeed, let $\mathbf{P}_\xi(E)$ be the probability attributed to the generic event E when the events $E_1, E_2, \ldots, E_n, \ldots$ are considered independent and equally probable with probability ξ; the probability $\mathbf{P}(E)$ of the same generic event, the E_i being exchangeable events with the limiting distribution $\Phi(\xi)$, is

$$\mathbf{P}(E) = \int_0^1 \mathbf{P}_\xi(E) \, d\Phi(\xi).^5 \tag{19}$$

(4)   I had then reserved the name "phenomenon" for the case of exchangeable trials; I now believe it preferable to use this word in the sense which is commonly given to it, and to specify, if it should be the case, that it is a question of a phenomenon whose trials are judged exchangeable.

(5)   It is clear that the particular case just mentioned—formula (18)—is obtained by putting E = "on n (given) trials, r results are favorable"; then, indeed,

$$\mathbf{P}_\xi(E) = \binom{n}{r} \xi^r (1-\xi)^{n-r}, \quad \mathbf{P}(E) = \omega_r^{(n)}.$$

This fact can be expressed by saying that the probability distributions **P** corresponding to the case of exchangeable events are linear combinations of the distributions P_ξ corresponding to the case of independent equiprobable events, the weights in the linear combination being expressed by $\Phi(\xi)$.

This conclusion exhibits an interesting fact which brings our case into agreement with a well known scheme, with which it even coincides from a formal point of view. If one has a phenomenon of exchangeable trials, and if Φ is the limiting distribution of the frequencies, a scheme can easily be imagined which gives for every problem concerning this phenomenon the same probabilities; it suffices to consider a random quantity X whose probability distribution is Φ and events which, conforming to the hypothesis $X = \xi$ (ξ any value between 0 and 1), are independent and have a probability $p = \xi$; the trials of a phenomenon constructed thus are always exchangeable events. Further on, we will analyze the meaning of this result, after having examined its extension to exchangeable random quantities. For the moment, we will limit ourselves to deducing the following result: in order that Φ may represent the limiting distribution corresponding to a class of exchangeable events, it is necessary and sufficient that the distribution be limited to values between 0 and 1 [so that $\Phi(-\epsilon) = 0$, $\Phi(1 + \epsilon) = 1$ when $\epsilon > 0$]; in other words it is necessary that the ω_h be the moments of a distribution taking values between 0 and 1, or again that $(-1)^s \Delta^s \omega_r \geqslant 0$ ($r, s = 1, 2, \ldots$), as results from the expression for $\omega_n^{(r)}$.

If only the probabilities of the various frequencies on n trials, $\omega_0^{(n)}$, $\omega_1^{(n)}$, $\omega_2^{(n)}$, $\cdots$, $\omega_n^{(n)}$, are known, the condition under which there can exist a phenomenon consisting of exchangeable trials for which the $\omega_r^{(n)}$ have the given values, will clearly be that the corresponding $\omega_1, \omega_2, \ldots, \omega_n$ be the first n moments of a distribution on $(0, 1)$; these ω_h can be calculated as a function of the $\omega_r^{(n)}$ by the formula

$$\omega_h = \sum_{r=h}^{n} \omega_r^{(n)} \frac{r! \, (n-h)!}{n! \, (r-h)!}; \tag{20}$$

finally, the condition that $\omega_1, \ldots, \omega_n$ be the first n moments of a distribution on $(0, 1)$ is that all the roots of the polynomial

$$f(\xi) = \begin{vmatrix} 1 & \xi & \xi^2 & \cdots & \xi^k \\ \omega_0 & \omega_1 & \omega_2 & \cdots & \omega_k \\ \omega_1 & \omega_2 & \omega_3 & \cdots & \omega_{k+1} \\ \cdots & \cdots & \cdots & \cdots & \cdots \\ \cdots & \cdots & \cdots & \cdots & \cdots \\ \omega_{k-1} & \omega_k & \omega_{k+1} & \cdots & \omega_{2k-1} \end{vmatrix} \quad \text{if } n = 2k - 1. \tag{21}$$

$$f(\xi) = \begin{vmatrix} 1 & \xi & \xi^2 & \cdots & \xi^k \\ \omega_1 & \omega_2 & \omega_3 & \cdots & \omega_{k+1} \\ \omega_2 & \omega_3 & \omega_4 & \cdots & \omega_{k+2} \\ \cdots & \cdots & \cdots & \cdots & \cdots \\ \cdots & \cdots & \cdots & \cdots & \cdots \\ \omega_k & \omega_{k+1} & \omega_{k+2} & \cdots & \omega_{2k} \end{vmatrix} \quad \text{if } n = 2k. \qquad (22)$$

fall in the interval $(0, 1)$, including the endpoints.[6]

CHAPTER IV

Exchangeable Random Quantities

Thus, as we have seen, in any problem at all concerning the exchangeable events $E_1, E_2, \ldots, E_n$, the probability will be completely determined either by the sequence of probabilities ω_n or by the limiting distribution of the frequency $\Phi(\xi)$ [or, what amounts to the same thing, by the corresponding characteristic function $\psi(t)$]. We have thus completely characterized the families of exchangeable events, and we have, in particular, elucidated the essential significance of $\Phi(\xi)$ connected with the fundamental result we have demonstrated: the probability distributions $\mathbf{P}$, corresponding to the case of exchangeability, are linear combinations of the probability distributions $\mathbf{P}_\xi$

(6)   This result follows from Castelnuovo [VII] (see also [VIII]), as we have noted in [29].

corresponding to the case of independence and equiprobability (probability = ξ). We have indeed,

$$P(E) = \int P_\xi(E)\, d\Phi(\xi) \qquad (19)$$

where $d\Phi(\xi)$ represents the distribution of weights in the linear combination.

We are going to extend this fundamental result to the case of exchangeable random quantities for which, up to now, we have only demonstrated the preliminary theorems, which we have used to establish certain results concerning the events themselves, rather than to solve the analogous problem, i.e. to characterize completely families of exchangeable random quantities.

Let us now consider the case of exchangeable random quantities and let us take an example to fix our ideas. In the study of exchangeable events, we have taken as an example the case of a game of heads or tails; let us now suppose that $X_1, X_2, \ldots, X_n$ represent measurements of the same magnitude; it suffices that the conditions under which the measurements are made do not present any apparent asymmetry which could justify an asymmetry in our evaluation of the probabilities, in order that we be able to consider them as exchangeable random quantities.

The extension of our earlier conclusions to this case will clearly be less easy than in the case of events, a *random quantity* being no longer characterized, from the probabilistic point of view, by a number (probability) as are the events, but by a function (for example, a distribution function or a characteristic function, etc.). Here the case of independence and equiprobability corresponds to the hypothesis of the independence of the random quantities X_i and the existence of a general distribution function $V(x)$; by calling $P_v(E)$ the probability attributable to a generic event E, when the X_i are considered to be independent and to have the same distribution function V, the linear combinations will be distributions of the type

$$P(E) = \Sigma\, c_i P_{v_i}(E)$$

(with the weights $c_i > 0$, $\Sigma c_i = 1$); in the limit

$$P(E) = \int P_v(E)\, d\,\mathscr{F}(V), \qquad (23)$$

the integral being extended over the function space of distribution functions, and the distribution of weights being characterized by the functional $\mathscr{F}(V)$, in a manner which will be made precise in what follows. Even before knowing the exact meaning of this integration, one is led to notice immediately that if P(E) is a linear combination of the $P_v(E)$ one has the case of exchangeability: it suffices to observe that each $P_v(E)$ giving the same probability to the events defined in a symmetrical[1] fashion in relation to $X_1, \ldots, X_n, \ldots$, the same condition will necessarily be satisfied by every linear combination P(E); it is a question then only of proving the inverse, i.e. of showing that, in the case of exchangeability, P(E) is necessarily of the form $\int P_v(E) \, d \, \mathcal{F}(V)$.[2]

The definition of the integral

$$\int f(V) \, d \, \mathcal{F}(V)$$

that we must introduce over the function space is only a completely natural generalization of the Stieltjes-Riemann integral:[3] by subdividing the space of distribution functions into a finite number of partial domains in any way whatever, we consider the expressions $\Sigma \, \overline{f_i} \, c_i$ and $\Sigma \, \underline{f_i} c_i$ where c_i is the weight of a generic element of these parts, and $\overline{f_i}$ and $\underline{f_i}$ are respectively the upper and lower bounds of the values taken by the function f in these domains. The lower bound of $\Sigma \, \overline{f_i} \, c_i$ and the upper bound of $\Sigma \, \underline{f_i} \, c_i$, when the subdivision is changed in all possible ways, are respectively the superior and inferior integral of f, extended to the function space of distribution functions in relation to the distribution of weights $\mathcal{F}$; when they coincide,

(1) Symmetric in the sense that, for example, the event E = "the point determined by the coordinates $X_1, X_2, \ldots, X_n$ will fall in the domain D" (in Euclidean space of n dimensions) is symmetrical to the events consisting in the same eventuality for one of the $n!$ points $(X_{i_1}, X_{i_2}, \ldots, X_{i_n})$ corresponding to the $n!$ permutations of the coordinates. In particular:

(for D rectangular):
$$E = \text{``}a_h < X_h < b_h \, (h = 1, 2, \ldots, n)\text{''}$$
and $$\text{``}a_h < X_{i_h} < b_h \, (h = 1, 2, \ldots, n)\text{''};$$

(for D spherical):
$$E = \text{``}\Sigma \, (X_h - a_h)^2 < \rho^2\text{''}$$
and $$\text{``}\Sigma \, (X_{i_h} - a_h)^2 < \rho^2\text{''};$$

(for D a half-space):
$$E = \text{``}\Sigma \, a_h X_h > a\text{''} \text{ and } \text{``}\Sigma \, a_h X_{i_h} > a,\text{''} \cdots$$

(2) One can accept this result and omit the following developments which are devoted to proving it and making it precise (toward the end of Chap. IV), without prejudice to an overall view of the thesis maintained in these lectures.

(3) For the reasons which make us regard the Stieltjes-Riemann integral as more appropriate to the calculus of probability, see [58] and [64].

their common value is precisely the integral $\int f(V)\,d\,\mathcal{F}(V)$ that we are going to examine more closely.

We are going to show that, in the circumstances that interest us, this integral exists, and that in order to determine its value, it suffices to know the weight for some very simple functional domains of distribution functions. Suppose to begin with the $f(V)$ depends only on the values

$$y_1 = V(x_1),\ y_2 = V(x_2),\ldots,\ y = V(x_s)$$

which the function V takes on a finite given set of abscissas $x_1, x_2, \ldots, x_s$; $f(V)$ is thus the probability that n random variables following the distribution V will all fall in a rectangular domain D, the first falling between x_1 and x_1', the second between x_2 and $x_2'\cdots$, the last between x_n and x_n'. This probability is[4]

$$f(V) = [V(x_1') - V(x_1)]\,[V(x_2') - V(x_2)] \cdots [V(x_n') - V(x_n)]$$
$$= (y_1' - y_1)(y_2' - y_2) \cdots (y_n' - y_n) \qquad (s = 2n) \qquad (24)$$

It is clear that in order to evaluate the integral of such a function, it is sufficient to know the weights of the functional domains defined only by the ordinates $y_1, \ldots, y_s$ corresponding to the abscissas $x_1, \ldots, x_s$, i.e. the weights of the domains of the space of s dimensions defined by $y_1, \ldots, y_s$; if f is a continuous function of the y_i it will suffice to know the weights of the domains defined by the inequalities $y_i < a_i\,(i = 1, 2, \ldots, s)$. The significance of these domains is the following: they comprise the distribution functions V whose representative curve $y = V(x)$ remains below each of the s points (x_i, a_i). Let $\Phi(x)$ be the stepwise curve of which the points (x_i, a_i) are the lower corners; the above condition can now be expressed by $V(x) < \Phi(x)$ [for all x],[5] and the weights of the set of distribution functions V such that $V(x) < \Phi(x)$ will be designated by $\mathcal{F}(\Phi)$; thus we give a concrete meaning to $\mathcal{F}$ which until now has represented a distribution of weights in a

(4) It is not necessary to be particularly concerned with the discontinuity points: indeed, a determinate function V is continuous almost everywhere (better: everywhere except, at most, on a denumerable set of points), and likewise in the integration, the weight of the set of distribution functions having x as a point of discontinuity is always zero, except, at most, for a denumerable set of points x; it suffices to observe that these points are the points of discontinuity of $\Phi(x) = \int V(x)\,d\,\mathcal{F}(V)$, and that $\Phi(x)$ is a distribution function.

(5) It is always understood that an inequality like $f(x) < g(x)$ between two functions means that it holds for all x (unless one has explicitly a particular case in mind when it is a question of a determinate value x).

purely symbolic way. In this case the integral $\int f(V)\,d\mathcal{F}(V)$ is only the ordinary Stieltjes-Riemann integral in the space of s dimensions. If $f(V)$ does not depend solely on the ordinates of $V(x)$ for a finite set of abscissas $x_1, \ldots, x_s$ we will consider the case where it is possible to approach $f(V)$, from above and below, by means of functions of the preceding type, in such a way that the value of the integral will be uniquely determined by the values approached from above and below. In other words, it will be necessary that, for an arbitrary ϵ, one be able to find two functions $f'(V)$ and $f''(V)$ depending on a finite number of values $V(x_i)$, such that

$$f'(V) \leqslant f(V) \leqslant f''(V) \quad \text{and} \quad \int f'(V)\,d\mathcal{F}(V) > \int f''(V)\,d\mathcal{F}(V) - \epsilon.$$

We return to the case of n independent random quantities having the distribution $V(x)$: if $f(V)$ is the probability that the point $(X_1, X_2, \ldots, X_n)$ falls in a domain D which is not reducible to a sum of rectangular domains, f' and f'' can represent the analogous probabilities for the domains D' contained in D, and D'' containing D, each formed from a sum of rectangular domains.

We have no need to pursue to the end the analysis of the conditions of integrability; we will content ourselves with having shown that they are satisfied in some sufficiently general conditions which contain all the interesting cases. We now return to the problem concerning the exchangeable random quantities $X_1, X_2, \ldots, X_n, \ldots$ in order to show the existence of the functional $\mathcal{F}$ having a meaning analogous to $\Phi(\xi)$ for exchangeable events. Let V be a stepwise function of which the lower corners are the s points

$$(x_i, y_i)\,(i = 1, 2, \ldots, s;\ x_{i+1} > x_i;\ y_{i+1} > y_i);$$

we will designate by $\mathcal{F}_h(V)$ the probability that, of h numbers $X_1, X_2, \ldots,$ X_h, hy_1 at the most will exceed x_1, hy_2 at the most will exceed $x_2, \ldots, hy_s$ at the most will exceed x_s, or, in other words, the probability

$$P\left\{G_h(x) \leqslant V(x)\right\}$$

that the distribution function $G_h(x)$ of the values of $X_1, X_2, \ldots, X_h$ never exceeds $V(x)$. More precisely, the function $G_h(x)$ is the "observed distribution function" resulting from the observation of $X_1, \ldots, X_h$; it represents the stepwise curve of which the ordinate is zero to the left of the smallest of the h numbers $X_1, \ldots, X_h$, equal to $1/h$ between the smallest and the second, equal afterwards to $2/h, 3/h, \ldots, (h-1)/h$, and finally equal to unity to the right of the largest of the h numbers considered. The steps of $G_h(x)$ are placed on the points of the axes of abscissas which correspond to the values X_i; before knowing these values, $G_h(x)$ is a random function, since these abscissas are random quantities.

It is easy to show, by extending a theorem given by Glivenko[6] for the case of independent random quantities to the case of exchangeable random quantities, that it is very probable that for h and k sufficiently large, $G_h(x)$ and $G_k(x)$ differ very little, and in, the case of a set of *successive* averages $G_h(x)$, $G_{h+1}(x)$, . . . , we have a strong stochastic convergence. By dividing the x axis into a sufficiently large finite number of points $x_1, \ldots, x_s$ the proof can be based on that given for the analogous properties in the case of exchangeable events. For a given x, $G_h(x)$ and $G_k(x)$ give respectively the frequencies Y_h and Y_k for the h and k trials of the set of exchangeable events $E_i = (X_i < x)$; the difference between $G_h(x)$ and $G_k(x)$ then has

standard deviation less than $\sqrt{\dfrac{1}{h} + \dfrac{1}{k}}$ [see formula (10)], and the proba-

bility that it exceeds ϵ can be made as small as one wishes by choosing h and k larger than a sufficiently large number N. By taking N so that the probability of a difference greater than ϵ is less than θ/s for each of the abscissas

$$x = x_1, x_2, \ldots, x_s,$$

we see that, except in a case whose total probability is less than θ, the two functions $G_h(x)$ and $G_k(x)$ will not differ by more than ϵ for any of the abscissas $x_1, \ldots, x_s$.

Under these conditions, the probability $\mathscr{F}_k(V - \epsilon)$ that $G_k(x)$ will not exceed the stepwise curve $V(x) - \epsilon$ for any x, which is to say the probability of having

$$G_k(x_i) < y_i - \epsilon \qquad (i = 1, 2, \ldots, s)$$

can not be more than $\mathscr{F}_h(V) + \theta$, for, in order to satisfy the imposed conditions, it is necessary either that $G_h(x)$ not exceed $V(x)$ for any x, or that we have $G_h(x) - G_k(x) > \epsilon$ for at least one of the abscissas $x_1 \cdots x_s$. We thus have

$$\mathscr{F}_k(V - \epsilon) - \theta \leqslant \mathscr{F}_h(V) \leqslant \mathscr{F}_k(V + \epsilon) + \theta \tag{25}$$

(the second inequality can be proved in the same way); by defining convergence in an appropriate way (as for distribution functions[7]), one concludes

(6)   [XIII], see also Kolmogorov [XVIII], and [45].
(7)   See Lévy [XX], p. 194.

that $\mathcal{F}_h \to \mathcal{F}$; it is the functional $\mathcal{F}$ which allows us to characterize the family of exchangeable random quantities we have in mind.

To prove the fundamental formula

$$P(E) = \int P_V(E) \, d\mathcal{F}(V) \tag{23}$$

we remark that we have, for all h,

$$P(E) = \int P_{h,V}(E) \, d\mathcal{F}(V) \tag{26}$$

where $P_{h,V}(E)$ is the probability of E, given the hypothesis

$$G_h(x) = V(x).$$

If the event E depends on the n first random quantities $X_1, \ldots, X_n$ (to fix our ideas by a simple example, let us imagine that the event E consists of X_1 falling between a_1 and b_1, X_2 between a_2 and b_2, \ldots, X_n between a_n and b_n), it will naturally be necessary to suppose $h \geqslant n$; if h is very large in relation to n, it is clear that $P_{h,V}(E) \cong P_V(E)$, for the probability $P_{h,V}(E)$ is obtained by supposing $X_1, \ldots, X_n$ chosen by chance, simultaneously (that is, without repetition) from among the h values where $G_h = V$ is discontinuous, whereas $P_V(E)$ is the analogous probability obtained by considering all the combinations possible on the supposition of independent choices. The fact of including or excluding repetitions has a more and more negligible influence as $h \to \infty$; thus $P_{h,V}(E) \to P_V(E)$. This relation and the relation $\mathcal{F}_h(V) \to \mathcal{F}(V)$ provide the proof that

$$P(E) = \int P_{h,V}(E) \, d\mathcal{F}_h(V) = \int P_V(E) \, d\mathcal{F}(V).$$

We shall consider a particular type of event E, which will permit us to analyze the relation between the functional distribution given by $\mathcal{F}$, relative to the exchangeable random quantities X_i, and the linear distributions $\Phi_x(\xi)$, that is to say, the limiting distributions $\Phi(\xi)$, related to the events $E_i = (X_i < x)$. An event E will belong to the particular type envisaged if it expresses a condition depending solely on the fact that certain random quantities $X_1, \ldots, X_n$ are less than or greater than[8] a unique given number x. For

(8)   This case of equality has a zero probability, except for particular values of x finite or at most denumerable in number; we neglect this case for simplicity, observing, besides, that it does not entail any essential difficulty, but only the consideration of two distinct values of the distribution function to the left and right of x.

example, E = "X_1, X_3, X_8 are $> x$, X_2 and X_7 are $< x$"; E = "among the numbers X_2, X_3, X_9, X_{12} there are three which are $> x$ and one $< x$"; E = "in the sequence $X_1 X_2 \cdots X_{100}$ there are no more than three consecu-tive numbers $> x$"; etc. In other words, the event E is a logical combination of the $E_i = (X_i < x)$ for a unique given x.

The theory of exchangeable events tells us that the probability of any event E of this type is completely determined by the knowledge of $\Phi_x(\xi)$, and we can express this probability with the aid of $\mathcal{F}(V)$; we can then express $\Phi_x(\xi)$ by means of $\mathcal{F}(V)$, and we then have precisely

$$\Phi_x(\xi) = \int_{V(x) < \xi} d\mathcal{F}(V); \quad d\Phi_x(\xi) = \int_{\xi < V(x) < \xi + d\xi} d\mathcal{F}(V). \quad (27)$$

Indeed, let $E^{(n)}$ be the event consisting in this; that the frequency of values $< x$ on the first n trials X_1, ..., X_n not exceed ξ; by definition $\Phi_x(\xi) = \lim P(E^{(n)})$, and moreover $P_V(E^{(n)})$, for n very large, is very close[9] to 0, if $V(x) > \xi$, or to 1, if $V(x) < \xi$; we will have, then,

$$\Phi_x(\xi) = \lim \int P_V(E^{(n)}) \, d\mathcal{F}(V)$$
$$(27)$$
$$= \int_{V(x) < \xi} 1 \cdot d\mathcal{F}(V) + \int_{V(x) > \xi} 0 \cdot d\mathcal{F}(V) = \int_{V(x) < \xi} d\mathcal{F}(V).$$

One can deduce from this (or, better, obtain directly) the following result: $\omega_r^{(n)}(x)$, the probability that r out of n random quantities X_{i_1}, ..., X_{i_n}, chosen in advance, should be $< x$, is equal to

$$\omega_r^{(n)}(x) = \binom{n}{r} \int [V(x)]^r [1 - V(x)]^{n-r} \, d\mathcal{F}(V), \quad (28)$$

and in particular for $r = n$:

$$\omega_n(x) = \int [V(x)]^n \, d\mathcal{F}(V), \quad \omega_1(x) = \int V(x) \, d\mathcal{F}(V); \quad (29)$$

this last formula giving, in particular, the probability that a general fixed number X_i is less than x; and this is the distribution function attributed to each of the X_i before beginning the trials.

Up to now, in $\Phi_x(\xi)$, $\omega_r^{(n)}(x)$, $\omega_n(x)$, we have considered x only as a parameter which determines the events $E_i = (X_i < x)$, but which does not vary; if, on the contrary, these expressions are considered as functions of x,

(9) We recall that $P_V(E^{(n)})$ is simply the probability of a frequency $< \xi$ on n independent trials, with constant probability $p = V(x)$, and therefore has the value

$$\sum_{v < \xi n} \binom{n}{v} p^v q^{n-v} [q = 1 - p = 1 - V(x)].$$

certain remarks can be made which throw a new light on them. Let us consider n of the given random quantities: $X_1, X_2, \ldots, X_n$; $\omega_n^{(n)}$ is the probability that none of these numbers exceed x, and thus constitutes the distribution function of the maximum value among $X_1, \ldots, X_n$; $\omega_n^{(n)}(x) + \omega_{n-1}^{(n)}(x)$ is in an analogous way the distribution function of the next-to-largest of the numbers $X_1, \ldots, X_n$ arranged in increasing order; $\omega_n^{(n)}(x) + \omega_{n-1}^{(n)}(x) + \cdots + \omega_r^{(n)}$ that of the rth; and finally $\omega_n^{(n)}(x) + \cdots + \omega_1^{(n)}(x) = 1 - \omega_0^{(n)}(x)$ that of the smallest of the X_i. As the identity

$$\omega_1^{(n)}(x) + 2\omega_2^{(n)}(x) + 3\omega_3^{(n)}(x) + \cdots + n\omega_n^{(n)}(x) = n\omega_1(x) \qquad (30)$$

shows, the average of these n distribution functions is $\omega_1(x)$, that is to say, the distribution function of any one whatever of the X_i: this fact is very natural, for, according to the definition of exchangeability, each number X_i has the same probability $1/n$ of being the smallest, or the second, ..., or the greatest, and, in general, all the permutations of the X_i have the same probability $1/n!$ of being disposed in increasing order (if there exists a probability different from zero that the n values are not distinct, the modifications to make in these statements are obvious).

These exists a close relation between the distribution functions of a random quantity of determinate rank and the function $\Phi_x(\xi)$: by definition, $\Phi_x(\xi)$ is the limiting value, for $n \to \infty$, of the probability that of n random quantities $X_1, \ldots, X_n$ there will be at most ξn which are $< x$; this probability is equal to $\sum_r \omega_r^{(n)}(x)$, the sum being extended over the indices $r < \xi n$.

But this sum is the distribution function of those of the numbers $X_1, \ldots, X_n$ which occupy the rank "whole part of ξn" where the random quantities are arranged in order of increasing magnitude: by considering ξ fixed, $\Phi_x(\xi)$ is (as a function of x) the distribution function of the number of rank $\cong \xi n$ on a very large number n of given random quantities.

It is easily seen that $\Phi_x(\xi)$ is a never-decreasing function of ξ and x, that $\Phi = 0$ if $\xi < 0$, and $\Phi = 1$ if $\xi > 1$ (Φ is thus defined substantially only on the interval $0 \leqslant \xi \leqslant 1$), and finally that $\Phi \to 0$ and $\Phi \to 1$ respectively for $x \to -\infty$ and $x \to +\infty$. Conversely, each function $\Phi_x(\xi)$ having these properties can be associated in an infinite number of ways with a probability distribution for exchangeable random quantities; one such function $\Phi_x(\xi)$ being given, one can always construct a distribution of weights $\mathscr{F}(V)$ in function space, such that formula (27) holds. The simplest way of doing this is the following: let $V_\lambda(x) = \xi$ be the explicit equation of the contour line $\Phi_x(\xi) = \lambda$, which represents, due to the properties of $\Phi_x(\xi)$, a distribution function, and defines the distribution $\mathscr{F}(V)$ by attributing the weights $\lambda' - \lambda (\lambda' > \lambda)$ to the set of $V(x)$ such that

$$V_\lambda(x) < V(x) < V_{\lambda'}(x)$$

for all x; in this way the integration in function space is reduced to a simple integral:

$$\int f(V)\, d\mathcal{F}(V) = \int_0^1 f(V_\lambda)\, d\lambda. \tag{31}$$

We have, for example,

$$\omega_n(x) = \int_0^1 [V_\lambda(x)]^n\, d\lambda = \int_0^1 \xi^n\, d\Phi_x(\xi); \tag{32}$$

this suffices to show that the distribution we have obtained satisfies the desired condition; it results directly from the calculation of

$$\Phi_x(\xi) = \int_{V(x)<\xi} d\mathcal{F}(V)$$

$$= \int_{V_\lambda(x)<\xi} d\lambda = \left\{ \text{the value of } \lambda \text{ for which } V_\lambda(x) = \xi \right\} \tag{33}$$

However, there always exists an infinity of other distributions $\mathcal{F}(V)$ corresponding to the same function $\Phi_x(\xi)$: it suffices to observe, for example, that if one puts in any way at all

$$\Phi_x(\xi) = c_1 \Phi_x^{(1)}(\xi) + c_2 \Phi_x^{(2)}(\xi) + \cdots + c_k \Phi_x^{(k)}(\xi)$$

with $c_i > 0$, $\Sigma\, c_i = 1$, the $\Phi_x^{(i)}(\xi)$ satisfying the same conditions as Φ, and if one introduces the corresponding $V_\lambda^{(i)}(x)$, one will always have

$$\omega_n(x) = \Sigma\, c_i \int_0^1 [V^{(i)}(x)]^n\, d\lambda. \tag{34}$$

The function $\Phi_x(\xi)$ thus characterizes neatly all the families of exchangeable events $E_i = (X_i < x)$ for any x whatever, but this does not suffice in problems where the interdependence of these various families comes into play: complete knowledge of the distribution $\mathcal{F}(V)$ in function space is then indispensable.

It should be noted once more that if one were to consider exchangeable random elements of any space whatever, one would arrive at perfectly analogous results: implicitly, we have already indeed considered some exchangeable random functions, since, for example, the $G_h(x)$, the distribution functions of $X_{i_1}, X_{i_2}, \ldots, X_{i_h}$ constitute a family of exchangeable random functions, when all possible groups $i_1, i_2, \ldots, i_h$ are considered. The general result which has been established for events and for random quantities, and which could be demonstrated for random elements of any space whatever, may be expressed by saying that *the probability distribution of classes of exchangeable random elements are "averages" of the probability distributions of classes of independent random elements.*[h]

CHAPTER V

Reflections on the Notion of Exchangeability

We have thus established the general notion of stochastic exchangeability, and obtained the fundamental result which permits the characterization of probability distributions corresponding to the case of exchangeability as some linear combination of the distributions corresponding to the case of independence and equiprobability. We will now try to show how this would be interpreted according to the current conceptions, to exhibit the reasons which keep us from holding such an opinion, and to explain the significance of the notion of exchangeability as well as the role that it is called to play, according to our point of view, in the calculus of probability.

(*h*)   The rigorous proof of this result under weak conditions on the space concerned has been the subject of several developments; for example, Hewitt, E., and Savage, L. J. (1955), "Symmetric measures on Cartesian products," *Trans. Amer. Math. Soc.,* **80**, 470-501 (and papers cited in the bibliography of this article), and, among later contributions, Bühlmann, H. (1960), "Austauschbare stochastische Variablen und ihre Grenzwertsätze," *Univ. Calif. Publ. Stat.,* **3**, 1-35, and Freedman, D.A. (1962), "Invariants under mixing which generalize de Finetti's theorem," *Ann. Mat. Stat.,* **33**, 916-923. An extension in a different sense (so as to include, for instance, cases of the type of Markov chains) has been dealt with by the present author in "Sur la condition d'équivalence partielle'" (Colloque Genève, 1937), *Act. Sci. Ind. No. 739,* Hermann, Paris, 1938, 5-18, and in "La probabilità e la statistica nei rapporti con l'induzione, secondo i diversi punti di vista" (Summer course at Varenna, 1959), *Induzione e Statistica,* Instituto Matematico dell'Università, Cremonese, Rome, 1959 (particularly, §8, pp. 92-100). An English translation of this last paper is almost ready and will probably be published as soon as some additional material can be prepared.

Let us consider, for example, an urn of unknown composition, and let us draw out balls, replacing each ball after it is drawn. The ratio of the number of white balls to the total number can have various possible values ρ_i, to which we attribute certain probabilities c_i. These drawings are exchangeable events, and the probability of a given outcome is $P(E) = \sum_i c_i P_i(E)$, where there is a probability $P_i(E)$ corresponding to the composition ρ_i; vice versa, as we have already remarked, if the E_i are exchangeable events, the distribution $P(E)$ is always of the form $P(E) = \sum_i c_i P_i(E)$ $(c_i > 0, \sum_i c_i = 1)$ or of the limiting form $P(E) = \int P_\xi(E) \, d\Phi(\xi)$ (Stieltjes' integral), where P_ξ is the probability distribution corresponding to the hypothesis of independence and constant probability ξ. The "exchangeable events" correspond then to those which we would ordinarily distinguish as "independent events with constant but unknown probability p," and $\Phi(\xi)$ would be "the probability that this 'unknown probability' p will be smaller than ξ." In a similar fashion, "exchangeable random quantities" correspond to those that we would call "independent random quantities with the constant but unknown probability distribution $G(x)$," and the functional $\mathscr{F}$ would give "the probability distribution of this unknown distribution," $\mathscr{F}(V)$ being the probability that $G(x) \leqslant V(x)$[1] (for V stepwise with a finite number of steps). But then, you say, why proceed from a novel definition, and make such efforts to conclude that it characterizes nothing else than this well-known case?

It is not without reason that we have considered ourselves obliged to proceed in this way. The old definition cannot, in fact, be stripped of its, so to speak, "metaphysical" character: one would be obliged to suppose that beyond the probability distribution corresponding to our judgment, there must be another, unknown, corresponding to something real, and that the different hypotheses about the unknown distribution—according to which the various trials would no longer be dependent, but independent— would constitute *events* whose probability one could consider. From our point of view these statements are completely devoid of sense, and no one has given them a justification which seems satisfactory, even in relation to a different point of view. If we consider the case of an urn whose composition is unknown, we can doubtless speak of the probability of different compositions and of probabilities relative to one such composition; indeed, the assertion that there are as many white balls as black balls in the urn expresses an objective fact which can be directly verified, and the conditional probability, relative to a given objective event, has been well defined. If, on the contrary, one plays heads or tails with a coin of irregular appearance, as in

(1)   See note (1), page 83.

the example of Chap. III, one does not have the right to consider as distinct hypotheses the suppositions that this imperfection has a more or less noticeable influence on the "unknown probability," for this "unknown probability" cannot be defined, and the hypotheses that one would like to introduce in this way have no objective meaning. The difference between these two cases is essential, and it cannot be neglected; one cannot "by analogy" recover in the second case the reasoning which was valid in the first case, for this reasoning no longer applies in the second case. If, after numerous drawings, the observed frequency of the white balls is f, why do we attribute a value close to f to the probability that the ball will be white in one of the drawings which is going to follow? It can be answered that after the observation of such a frequency we attribute a very large value to the probability that the number of white balls will come very close to the fraction f of the total, and further, by supposing this fraction to be ρ, we judge that the drawings are independent and have all the same probability $p = \rho$. This explanation is perfectly satisfactory even from the subjectivistic point of view, and does not differ formally from that which is ordinarily given and which reduces, finally, to the theorem of compound probabilities. But in the preceding case of heads or tails, it is otherwise: the corresponding terms which would allow analogous reasoning do not exist. If, nevertheless, we want to reason in an identical and rigorous way in the two cases, it is necessary to begin by looking for the common elements which characterize them, and for those elements which differentiate them.

The result at which we have arrived gives us the looked-for answer, which is very simple and very satisfactory: the nebulous and unsatisfactory definition of "independent events with fixed but unknown probability" should be replaced by that of "exchangeable events." This answer furnishes a condition which applies directly to the evaluations of the probabilities of individual events and does not run up against any of the difficulties that the subjectivistic considerations propose to eliminate. It constitutes a very natural and very clear condition, of a purely qualitative character, reducing to the demand that certain events be judged equally probable, or, more precisely, that all the combinations of n events $E_{i_1}, \ldots, E_{i_n}$ have always the same probability, whatever be the choice or the order of the E_i. The same simple condition of "symmetry" in relation to our judgments of probability defines exchangeable random quantities, and can define, in general, exchangeable random elements in any space whatever. It leads in all cases to the same practical conclusion: a rich enough experience leads us always to consider as probable future frequencies or distributions close to those which have been observed.

Following the demonstration of the existence of a limiting distribution, this fact can be explained by reasoning almost parallel to that which one ordinarily employs when one takes account of the "unknown probability," but which does not give rise to the same criticisms. For the "unknown probability" we substitute the frequency on the first N trials, with N large enough so that the corresponding probability distribution coincides practically with the limiting distribution: $\Phi_N(\xi) \cong \Phi(\xi)$, and so that the number of trials with which one is concerned is negligible in relation to N; thus $\binom{R}{r}\binom{S}{s} : \binom{N}{n}$ (the probability that on n trials chosen at random among the N of which R are favorable and S unfavorable there are r favorable and s unfavorable) is practically equal to $\binom{n}{r} \xi^r (1 - \xi)^s$ with $\xi = \frac{R}{N}$, $1 - \xi = \frac{S}{N}$. One can then reason as follows: consider as possible hypotheses the N + 1 possible frequencies on N trials, their probabilities being the $\omega_h^{(N)}$ ($h = 0$, $1, \ldots,$ N), and observe that the hypotheses for which R/N is close to r/n are precisely those according to which the probability of a frequency r/n on n trials is the strongest, and in consequence,[2] those for which the probability conditioned on the observation of the frequency r/n on n trials is the most strongly augmented in relation to the unconditional probability. We conclude finally that the hypotheses closest to the observed result take on a more and more preponderate importance when the number of observations increases, and that this leads us necessarily to make our prediction approach the observation. More precisely, we demonstrate that the limiting distribution $\overline{\Phi}$, given the observation of the frequency r/n on n trials, is such that

$$d\overline{\Phi}(\xi) = \alpha\xi^r(1 - \xi)^s \, d\Phi(\xi) \; [x \text{ such that } \int_0^1 \alpha\xi^r(1 - \xi)^s \, d\Phi = 1] \qquad (35)$$

and that the corresponding characteristic function is

$$\overline{\psi}(t) = \alpha D^r(i - D)^s \, \psi(t) \; [x \text{ such that } \overline{\psi}(0) = 1; D = d/dt; i = \sqrt{-1}\,]. \qquad (36)$$

In particular, the probability on an additional trial, relative to this hypothesis, will be

$$p_r^{(n)} = \int \xi \, d\overline{\Phi} = \int \xi \cdot \alpha\xi^r(1 - \xi)^s \, d\Phi; \qquad (37)$$

that is to say, the mean of the ξ from (0,1) with the weights $\alpha\xi^r(1 - \xi)^s \, d\Phi$ in place of the weights $d\Phi$; the ξ around the maximum $\xi = r/n$ of $\xi^r(1 - \xi)^s$ are evidently strengthened more and more.

(2)   See formula (2), p. 70, and the relevant explanation.

As a function of the $\omega_h^{(n)}$ we have easily

$$p_r^{(n)} = \frac{r+1}{n+1} \cdot \frac{\omega_{r+1}^{(n+1)}}{\omega_r^{(n)}} = \frac{r+1}{n+2+(s+1)\left(\dfrac{\omega_r^{(n+1)}}{\omega_{r+1}^{(n+1)}} - 1\right)} ; \qquad (38)$$

it is interesting to see that this formula—which already explains, though incompletely, the influence of the observed frequency on the evaluation of the probability—can be derived in a direct and very elementary way from the definition of "exchangeable events." From this definition it can indeed be inferred that

$$\frac{\omega_r^{(n)}}{\binom{n}{r}} = \frac{\omega_r^{(n+1)}}{\binom{n+1}{r}} + \frac{\omega_{r+1}^{(n+1)}}{\binom{n+1}{r+1}}, \qquad (39)$$

for the first member expresses the probability that r given events on the first n trials are favorable, and the second gives the sum of the probabilities that the said combinations will occur with, respectively, a favorable or unfavorable result on the $n + 1$'st trial. Simplifying, we have

$$\omega_r^{(n)} = \frac{s+1}{n+1} \, \omega_r^{(n+1)} + \frac{r+1}{n+1} \, \omega_{r+1}^{(n+1)} \qquad (40)$$

and with the help of this identity we obtain

$$
\begin{aligned}
p_r^{(n)} &= \frac{\omega_{r+1}^{(n+1)}}{\binom{n+1}{r+1}} : \frac{\omega_r^{(n)}}{\binom{n}{r}} = \frac{r+1}{n+1} \frac{\omega_{r+1}^{(n+1)}}{\omega_r^{(n)}} \\
&= \frac{(r+1)\,\omega_{r+1}^{(n+1)}}{(s+1)\,\omega_r^{(n+1)} + (r+1)\,\omega_{r+1}^{(n+1)}} \\
&= \frac{r+1}{n+2+(s+1)\left(\dfrac{\omega_r^{(n+1)}}{\omega_{r+1}^{(n+1)}} - 1\right)} \qquad \text{Q.E.D.} \quad (38)
\end{aligned}
$$

This formula acquires a particular significance in Laplace's case where the $\omega_r^{(n)}$ do not depend on r, and where one has $\omega_r^{(n)} = \dfrac{1}{n+1}$ (it can be verified immediately that this hypothesis is admissible, and that it corresponds to a homogeneous limiting law: $\Phi(\xi) = \xi$, $d\Phi(\xi) = d\xi$, for $0 \leqslant \xi \leqslant 1$).

In the case of Laplace one has simply, as is well known, $p_r^{(n)} = \dfrac{r+1}{n+2}$, for the other term in the denominator vanishes. The result is still the same if $\omega_{r+1}^{(n+1)} = \omega_r^{(n+1)}$; if on the contrary $\omega_{r+1}^{(n+1)}$ is greater or smaller than $\omega_r^{(n+1)}$, the probability $p_r^{(n)}$ will be respectively greater or smaller than in the Laplacean case. In any case, $p_r^{(n)}$ is close to $\dfrac{r+1}{n+2}$, and hence close to the frequency r/n, if the ratio differs little from unity; it thus suffices to admit this condition in order to justify easily the influence of observation on prediction in the case of exchangeable events.

With regard to exchangeable random quantities, or any exchangeable elements at all, the results and the demonstrations would be perfectly analogous; for these, as for events, the subjectivistic theory solves the problem of induction completely in the case of exchangeability, corresponding to the case which is most usually considered, and leads to the same conclusions generally admitted or demonstrated by means of vague and imprecise reasoning.

Every hypothesis, however, can be studied in the same style and by the same procedure. One cannot exclude completely *a priori* the influence of the order of events, and in consequence the attribution of probabilities differing more or less among themselves to the $\binom{n}{r}$ various combinations of r favorable results on the $n = r + s$ first trials. There would then be a number of degrees of freedom and much more complication, but nothing would be changed in the setting up and the conception of the problem, which would remain that presented at the beginning of Chap. III, before we restricted our demonstration to the case of exchangeable events, and which is essentially condensed in formula (4).

The influence of the order on the evaluation of the probabilities of A and $A \cdot E_{n+1}$ (*see* p. 79) does not modify, indeed, the way in which the problem is posed and answered according to the subjectivistic conception; one will only be led, in the general case, to take account of the circumstances which

106 BRUNO DE FINETTI

in the case of exchangeability are (by definition) neglected. One can indeed take account not only of the observed frequency, but also of regularities or tendencies toward certain regularities which the observations can reveal. Suppose, for example, that the n first trials give alternately a favorable result and an unfavorable result. In the case of exchangeability, our prediction for the following trial will be the same after these n trials as after any other experience of the same frequency of $\frac{1}{2}$,[3] but with a completely irregular sequence of different results; it is indeed the absence of any influence of the order on the judgments of a certain individual which characterizes, by definition, the events that he will consider "exchangeable." In the case where the events are not conceived of as exchangeable, we will, on the other hand, be led to modify our predictions in a very different way after n trials of alternating results than after n irregularly disposed trials having the same frequency of $\frac{1}{2}$; the most natural attitude will consist in predicting that the next trial will have a great probability of presenting a result opposite to that of the preceding trial.

It would doubtless be possible and interesting to study this influence of order in some simple hypotheses, by some more or less extensive generalization of the case of exchangeability, and some developments tied up with that generalization,[4] but this study is still to be done. What is essential in connection with our conception—and it is this that we want to insist on somewhat more—we have already learned through the theory of exchangeable events. Whatever be the influence of observation on predictions of the future, it never implies and never signifies that we *correct* the primitive evaluation of the probability $P(E_{n+1})$ after it has been *disproved* by experience and substitute for it another $P^*(E_{n+1})$ which *conforms* that that experience and is therefore probably *closer to the real probability*; on the contrary, it manifests itself solely in the sense that when experience teaches us the result A on the first n trials, our judgment will be expressed by the probability $P(E_{n+1})$ no longer, but by the probability $P(E_{n+1}|A)$, i.e. that which our initial opinion would already attribute to the event E_{n+1} considered as conditioned on the outcome A. Nothing of this initial opinion is repudiated or corrected; it is not the function P which has been modified (replaced by another P^*), but rather the argument E_{n+1} which has been replaced by $E_{n+1}|A$, and this is just to remain faithful to our original opinion (as manifested in the choice

(3)   In order that this should be exact, we suppose that n is even.
(4)   One could in the first place consider the case of classes of events which can be grouped into Markov "chains" of order $1, 2, \ldots, m, \ldots$, in the same way in which classes of exchangeable events can be related to classes of equiprobable and independent events.

of the function **P**) and coherent in our judgment that our predictions vary when a change takes place in the known circumstances.

In the same way, someone who has the number 2374 in a lottery with 10,000 tickets will attribute at first a probability of 1/10,000 to winning the first prize, but will evaluate the probability successively as 1/1000, 1/100, 1/10, 0, when he witnesses the extraction of the successive chips which give, for example, the number 2379. At each instant his judgment is perfectly coherent, and he has no reason to say at each drawing that the preceding evaluation of probability was not right (at the time when it was made). In the last analysis, each evaluation of probability different from 0 and 1 will surely be abandoned, for a well-determined event can only happen or not happen; an evaluation of probability only makes sense when and as long as an individual does not know the result of the envisaged event; given that he does not know this result (and therefore that he is not led to the definitive value 0 or 1), he can take account of successively more circumstances which would modify his judgment, in one sense or another, without its being a question of correction or rejection. It is in just the same way that we envisage the influence of observation on prediction in the general case of judgments founded on experience.

It is thus that when the subjectivistic point of view is adopted, the problem of induction receives an answer which is naturally subjective but in itself perfectly logical, while on the other hand, when one pretends to *eliminate* the subjective factors one succeeds only in *hiding* them (that is, at least, in my opinion), more or less skillfully, but never in avoiding a gap in logic. It is true that in many cases—as for example on the hypothesis of exchangeability—these subjective factors never have too pronounced an influence, provided that the experience be rich enough; this circumstance is very important, for it explains how in certain conditions more or less close agreement between the predictions of different individuals is produced,[5] but it also shows that discordant opinions are always legitimate. This does not make any change in the purely subjective character of the whole theory of probability.

We will return in the next and last chapter to these very general questions of principle, after a review of everything we have said so far, and some additional matters which throw more light on their *raison d'être* and scope.

---

(5)   It is the same point of view on which Poincaré several times insisted (and which inspired his well-known examples from roulette, shuffling cards, the distribution of small planets, etc.) (see, for example, [XVIII]); the only difference is in fact that we do not admit the conception that our initial opinion concerns "unknown distributions."

CHAPTER VI

Observation and Foresight

The need for clarity in scientific and philosophical thought has never appeared to be so essential as today: the most extensive critical analysis of the clearest intuitive concepts can no longer be considered a game for sophists, but is one of the questions which touch most directly on the progress of science. With each of our assertions, a question invariably surges into our mind: has this assertion really any meaning? To give only one example, we know that the notion of simultaneity seemed, not very long ago, perfectly clear and sure, to the point that it had been thought possible to consider time as a notion given *a priori*. Why do we no longer believe this today? Because we have been taught the necessity of conceiving of every notion from a point of view which can be called "operational".[1] Every notion is only a word without meaning so long as it is not known how to verify practically any statement at all where this notion comes up; in the example given above, this practical verification is furnished us by Einstein's procedure employing light signals. An analogous evolution took place some time ago in the mathematical sciences: once, for example, the problem of knowing if $1 - 1 + 1 - 1 + \cdots = \frac{1}{2}$ or not was considered in a nebulous, mysterious, metaphysical way; it sufficed to define what was to be understood by "limit" (for example ordinary limit, limit in the sense of Cesaro) and all the obscurities vanished.

It is perfectly natural that this need for clarity is felt deeply in the domain of probability, whether because this notion is very interesting from the mathematical point of view as well as from the experimental point of view, or whether because it seems recalcitrant to all attempts to make it precise. In that which concerns the mathematical side of the question, opinions do not seem to differ too much: formally,[2] the theory of probability is the theory of additive and non-negative functions of events; opinions diverge only on one point, the question of whether these functions need to be simply or completely additive (that is to say, additive only on finite sets, or

(1)   See, for example, Bridgman [II], and particularly the paragraphs *The Operational Character of Concepts* and *General Comments on the Operational Point of View* (p. 3-33).

(2)   See, for example, Cantelli [IV], [V], Kolmogorov [XVII], Lomnicki [XXI], etc.

also on denumerable sets). However, the truly essential aspect of the problem is naturally the question of the meaning and the value of the notion of probability, and on this terrain opinions differ strongly. Two completely opposed points of view are possible: the first, the most commonly accepted, considers the subjective element of the naïve notion of probability which is found in our everyday life as a dangerous element which ought to be eliminated in order that the notion of probability be able to attain a truly scientific status; the opposite point of view considers, on the contrary, that the subjective elements are essential, and cannot be eliminated without depriving the notion and theory of probability of all reason for existing. The difference between the two points of view is also very sharp from the philosophical point of view: according to the one, probability is an element which partakes of the physical world and exists outside of us; according to the other, it only expresses the opinion of an individual and cannot have meaning except in relation to him.

Both of these two points of view seek to give a well-defined meaning to probability statements, but the domains in which these concepts should receive a meaning are completely different. To give a verifiable meaning to probability statements in the external world would be to consider them not as something genuinely new, but as particular statements concerning the physical world, for example, as statements about the limits of certain frequencies. However, if one wanted to interpret the requirements of the operational point of view only within the framework of the external world, in a way which could be called positivistic, I think that the goal of making all our ideas clear could never be completely attained. We are sometimes led to make a judgment which has a purely subjective meaning, and this is perfectly legitimate; but if one seeks to replace it afterward by something objective, one does not make progress, but only an error. Rather than by seeking to bring everything back to the objective, one can attain clarity by reducing any such concept systematically to the subjective; the value of a concept would then result from the analysis of the deep and essential reasons which have made us, perhaps unconsciously, introduce it, and which furnish us with the explanation of its usefulness.

This point accepted, it should not be difficult to see that the definition based on the frequency limit[3] is far from clarifying the notion of probability; indeed, even if such a definition is accepted, one will not employ the calculus of probability with the object of knowing certain values of limits of frequencies; the object of employing the calculus will always be

(3)   See, for example, von Mises [XXV], [XXVI], Reichenbach [XXIX], [XXX], [XXXI]; see also Dörge [IX], where this conception has been modified according to criticisms which are also justified from our point of view.

that of judging as more or less likely the occurrence of certain facts, more or less complex, but verifiable in a finite time. These are the only events that interest us, and in regard to them the stated definition teaches us nothing. We only apply the notion of probability in order to make likely predictions: if I want to justify by the practical observation of a frequency the conviction that a neighboring frequency will appear in a certain group of *subsequent* trials, and if, for that, I proceed by estimating to begin with that the limiting frequency will be close to the observed frequency, and then that it is reasonable to expect a frequency close to the limiting frequency, I only introduce a mysterious intermediate notion through which the premises and conclusion are related indirectly by two subjective judgments in place of being related directly by a single subjective judgment [62], [63]. It does not help me at all to give the name of probability to the limiting frequency, or to any other objective entity, if the connection between these considerations and the subjective judgments which depend on them remains subjective. It is worth more, then, to seek to analyze directly the subjective element to which the notion of probability is directly anchored: it is this road that I have followed.

I know very well the doubts which are raised currently concerning this point of view, and it is for this reason that I propose to express, as clearly as it is possible for me, the way in which the problems for which the ordinary objections assume the most striking form are to be conceived and set up according to the subjectivistic conception.

There are three essential objections: it is doubted that the subjectivistic conception permits the definition of probability, the demonstration of the logical laws which govern it, and finally the explanation and justification of the applications that are made of it to the most unlike problems. Let us review rapidly our answers to these three objections.

The definition of probability that we have given is entirely irreproachable from the operational point of view, provided one admits that the latter is equally applicable in the psychological domain. The scheme of bets gives in principle a method of direct experimental measure of the degree of doubt relative to a given event. If the practical application sometimes runs up against an indeterminateness of this subjective degree of doubt which is to be measured, that is only a consequence of that limited degree of idealization without which it would always be impossible to attain precision and to develop any theory at all. The indeterminateness is doubtless stronger here than in the physical sciences because of the fact that the magnitude measured is subjective, but the difference is not essential; another definition, equally subjective and very similar, that it may perhaps be useful to compare with the subjective definition of probability, is that of the "utility" of Vilfredo Pareto [XXVII], who, in making it follow from "indifference curves," applied the operational point of view to psychological facts with perspicacity.

The fact that a direct estimation is not always possible constitutes the reason for the utility of the logical rules of probability: their practical end is to relate an evaluation, itself not very directly accessible, to others by means of which the determination of the first evaluation is made easier and more precise. By adopting the subjectivistic definition, these logical rules follow with rigor and ease from a single and very natural condition, that of *coherence,* which obliges us to take care in evaluating probabilities not to allow an adversary who bets against us the possibility of winning with certainty, whatever be the event that occurs, by a judicious combination of his stakes on the various events. The fundamental theorems (total probability, compound probability) are only the immediate corollaries of this fundamental condition. It can be seen that one could even eliminate everything quantitative, whether in the condition of coherence or in the definition of probability, in order to keep only the purely qualitative aspect of the definition (inequality between two probabilities) and of the condition of coherence (a small number of very simple axioms). The application of these logical rules in every case reduces to distinguishing whether, the probabilities being evaluated arbitrarily (but satisfying the condition of coherence) for the events of a certain class $\mathcal{E}$, the probabilities of other events are univocally determined by the condition of coherence, or whether there exists a limitation, or, finally, whether any values at all remain admissible.

From the logical point of view, the theory of probability would be only a polyvalent logic with a continuous scale of modalities,[4] superimposed on a logic of two values. This is to say that, for each event, one admits only two possible *results* (three for "conditional events," but that has only a formal significance); the infinity of intermediate modalities does not stem from an insufficiency of the logic of two values in this respect, but only serves to measure our doubt when we do not yet know which of the two objective modalities is correct.

The subjectivistic explanation of the most important applications of the calculus of probabilities constitutes a very delicate problem. It would not be difficult to admit that the subjectivistic explication is the only one applicable in the case of practical predictions (sporting results, meteorological facts, political events, etc.) which are not ordinarily placed in the framework of the theory of probability, even in its broadest interpretation. On the other hand it will be more difficult to agree that this same explanation actually supplies rationale for the more scientific and profound value that is attributed

(4) [62]; according to the opinions of Lukasiewicz [XXII], Mazurkiewicz [XXIII], and Reichenbach [XXIX]; it would also be a question of a logic of a continuous scale of modalities, which would not, however, be conceived as superimposed on a logic of two values. Some criticisms, to which this point of view lends itself, are developed in Hosiasson [XIV].

to the notion of probability in certain classical domains, and doubts will be expressed about the possibility that it offers of unifying the various conceptions of probability, appropriate to various domains, that until now it has been thought necessary to introduce. Our point of view remains in all cases the same: *to show that there are rather profound psychological reasons which make the exact or approximate agreement that is observed between the opinions of different individuals very natural, but that there are no reasons, rational, positive, or metaphysical, that can give this fact any meaning beyond that of a simple agreement of subjective opinions.*

The case of games of chance leads only to the observation of how the character of symmetry presented by the various "possible cases" can force us to judge them equally possible, but not to *impose* such an evaluation of probability logically. The frequency case, on the other hand, requires an elaborate analysis, which leads us to some fairly extended mathematical developments. Similar reasoning, hardly more complicated than that concerning games of chance, suffices to explain the dependence between the evaluation of probabilities of certain events and the prediction of the number among them that will occur—that is to say, the bond between the evaluation of probabilities and the prediction of frequencies. The essential question, and the only one which is a little less elementary, is the justification and the explanation of the reasons for which in the prediction of a frequency one is generally guided, or at least influenced, by the observation of past frequencies. It is a question of showing that there is no need to admit, as it is currently held, that *the probability of a phenomenon has a determinate value* and that it suffices to get to know it. On the contrary, the question can be posed in a way which has a perfectly clear sense from the subjectivistic point of view, by distinguishing on the one hand the probability of a trial considered as isolated, and on the other the probability of the same trial preceded by some others of which the result is supposed (by hypothesis) to be known.

We have studied the case where, in the evaluation of conditional probabilities, one is influenced only by the observed frequency. This case can be characterized in an equivalent, but simpler and more intuitive way, as the case where the several trials of the phenomenon considered—or, in general, the event considered—play a symmetric role in relation to any problem of probability; or better yet, as the case where the probability that a given r trials have a favorable result and s others an unfavorable result depends only on r and $s;$ or finally, as the case where the probability that n trials all have a favorable result is the same however the n-tuple is chosen. These conditions, which define "exchangeable events," have an immediate and very clear meaning from the point of view of the subjectivistic theory of probability, and there are numerous practical cases where they present themselves spontaneously to our minds. This suffices to explain our belief in the stability of the frequency, for, on this hypothesis, the probability of a subsequent trial, relative to the observation of a certain frequency, tends to coincide with the value of the latter. There is, however, a particular case, that of independence, for which the influence of past observations is rigorously zero. This case constitutes an exception; in all the other cases the influence of the acquired results tends to predominate when the number of observed cases is increased, though naturally in a way which is not uniform (one can have, for example, evaluations close to those which correspond to the case of independence for which this influence is zero). This *subjective* evaluation thus plays an essential role; the condition of "exchangeability" itself has, from the beginning, only a *subjective* value.

This reasoning is not applicable only to frequencies: in the case of exchangeable random quantities (and in general in the case of random elements of any space at all) a certain stability in the distribution of values can be justified in the same way and with similar reservations. Even the problem of smoothing a curve should be studied from this point of view:[5] the adjusted distribution curve would then be the probability distribution, conditioned on the observation of the values actually observed. This curve would depend on a subjective opinion, but to a smaller degree the richer the experience; on the other hand, one can see in this conception the true reason for the procedure of smoothing, and the conditions of "regularity" and "closeness to observed behavior" which suggest it. From this point of view, the conditions are no longer arbitrary or formal conditions, but the consequences of theorems on exchangeable random quantities and of the natural tendencies of our minds in the evaluation of probabilities of different kinds.

What I have said and shown for the case of exchangeability can clearly be repeated, with the necessary modifications, for less simple and less typical conditions which have been merely alluded to (end of Chap. V). The meaning of these conclusions is always the same: observation cannot confirm or refute an opinion, which is and cannot be other than an opinion and thus neither true nor false; observation can only give us information which is capable of *influencing* our opinion. The meaning of this statement is very precise: it means that to the probability of a fact conditioned on this information—a probability very distinct from that of the same fact not conditioned on anything else—we can indeed attribute a different value.

Thus, I think I have succeeded, if not in persuading those who are far from accepting the subjectivistic point of view, at least in proving that this point of view gives an irreproachable answer to all the usual questions, and that it permits their combination into a single coherent conception.

(5)   [56]; the point of view of Poincaré [XXVIII, pp. 204-206] is very similar.

Certain minds—convinced in other respects that the subjectivistic theory of probability constitutes a coherent conception, complete and perfectly acceptable in itself—will refuse to rally to it for reasons of a philosophical sort. One might indeed think that scientific concepts ought always to have a real meaning, that science ought to occupy itself exclusively with realities, and that the subjectivistic point of view would lead further away each day from this principle through the more and more extended application of probability to the physical sciences; not only that particular branch of mathematics which constitutes the calculus of probability, not only its applications to games and statistics, but also a greater part each day of the concepts of physics would cease to correspond to an objective reality. And one might say that the deterministic laws of the classical type and the statistical laws which have been substituted[6] for them would no longer have even that common, essential characteristic which has bound them together until now, namely, the connection with reality. Would there not then be an uncrossable abyss separating the two types of laws which coexist today in physics?*

In order to bridge this abyss, the point of view adopted here leads us in a completely natural way to a solution which is exactly the opposite of that which we habitually envisage: in place of extending the character of reality of the classical laws to the probability laws, we can try on the contrary to make even the classical laws participate in the subjective character of the statistical laws. I have already cited this sentence of Poincaré: "However solidly based a prediction may seem to us, we are never absolutely sure that experience will not refute it." Laws only have value for us, in that—and only in the sense that—we estimate it *very improbable,* after experience and the scientific analysis of its results, that a "law" should be disproved by the occurrence of an event contradicting a result that it had predicted. Rigid laws are only *proven* by experience in the sense of the verification of an agreement between them and a certain number of facts. To ask if these facts occur *because the law is true,* or to ask if the true law is not *different from that which we have,* with which it coincides only in these particular cases, or finally to ask if the law *does not exist,* are questions which from the operational point of view have no sort of meaning. There is always an infinity of explanations possible for the same group of observations; if we choose one of them, and if we state a law, it can only be for subjective reasons that make us consider it worthy of confidence. Rigid laws are formulated and accepted by our minds for the same reasons that lead us to formulate and accept any judgment of probability whatever; the only difference consists in the very high probability that we attribute, in the case of rigid laws, to their exact agreement with experimental facts. The probability is so high

(6)   For certain aspects of our opinions on this question, see also [36].

that we can call it "practically absolute certainty," or, simply "certainty," understanding all the while the qualification that is essential from the philosophical and logical point of view.

The notion of "cause" thus depends on the notion of probability, and it follows also from the same subjective source as do all judgments of probability [32]: this explanation seems to constitute the true logical translation of the conception of "cause" advanced by David Hume, which I consider the highest peak that has been reached by philosophy. The subjectivistic theory of probability will thus be able to open the field of science to this conception, whose significance and value seem not to have been sufficiently understood nor appreciated until now.

It is for these reasons that the theory of probability ought not to be considered an auxiliary theory for the branches of science which have not yet discovered the deterministic mechanism that "must" exist; instead it ought to be regarded as constituting the logical premises of all reasoning by induction. Just as the ordinary logic of two values is the necessary instrument of all reasoning where only the fact that an event happens or does not happen enters in, so the logic of the probable, the logic of a continuous scale of values, is the necessary instrument of all reasoning into which enters, visible or concealed, a degree of doubt, a judgment of practical certainty or practical impossibility, or finally, an estimation of the likelihood of any event whatever. Everything that does not reduce to a simple statement, to an isolated historical truth, everything that councils us for the future, even the belief that in leaving our room we will see as on other days the same streets and the same houses in their same places, all that constitutes a judgment of probability which is based, perhaps unconsciously and indistinctly, on the principles of the calculus of probability. This calculus thus constitutes the foundation of the greatest part of our thought, and we can well repeat with Poincaré, "Without it, Science would be impossible."

BIBLIOGRAPHY

[I]	Bertrand (J.)	*Calcul des probabilités,* Paris, 1889.
[II]	Bridgman (P.)	*The Logic of Modern Physics,* New York, 1927.
[III]	Cantelli (F. P.)	"Sulla probabilità come limite della frequenza," *Rendiconti Reale Accademia Nazionale dei Lincei,* Series 5, Vol. XXVI, 1st Semester (1917).
[IV]	------	"Una teoria astratta del calcolo della probabilità," *Giornale dell'Istituto Italiano degli Attuari,* Vol. III-2 (1932).
[V]	------	"Considérations sur la convergence dans le calcul des probabilités," *Annales de l'Institut Henri Poincaré,* Vol. V, fasc. 1 (1935).

[VI]	Castelnuovo (G.)	*Calcolo delle probabilità*, Zanichelli, Bologna, 1925.
[VII]	———	"Sul problema dei momenti," *Gior. Ist. Ital. Attuari*, Vol. I-2 (1930).
[VIII]	———	"Sur quelques problèmes se rattachant au calcul des probabilités," *Ann. Inst. H. Poincaré*, Vol. III, fasc. 4 (1933).
[IX]	Dörge (K.)	"Ueber das Anwendung der Wahrscheinlichkeitsrechnung und das Induktionsproblem," *Deutsche Mathematik*, **1** (1936).
[X], [XI]	Fréchet (R. M.)	"Sur l'extension du théorème des probabilités totales au cas d'une suite infinie d'événements," *Rendiconti Reale Istituto Lombardo di Scienze Lettere ed Arte*, Series 2, Vol. LXIII, first note, fasc. 11-15, second note, fasc. 16-18 (1930).
[XII]	Fréchet (R. M.) and Halbwachs (M.)	*Le calcul des probabilités à la portée de tous*, Dunod, Paris, 1925.
[XIII]	Glivenko (V.)	"Sulla determinazione empirica delle leggi di probabilità," *Gior. Ist. Ital. Attuari*, Vol. IV-1 (1933).
[XIV]	Hosiasson (J.)	"La théorie des probabilités est-elle une logique généralisée?" Vol. IV of *Actes du Congrès International de Philosophie Scientifique* (1935), Act. Scient. Indust. #391, Hermann, Paris, 1936.
[XV]	Khinchin (A.)	"Sur les classes d'événements équivalentes," *Mathematičeskii Sbornik, Recueil. Math. Moscou*, 39-3 (1932).
[XVI]	———	"Remarques sur les suites d'événements obéissant à la loi des grandes nombres," *Ibid.*
[XVII]	Kolmogorov (A.)	*Foundations of the Theory of Probability*, 2nd English edition (translation edited by N. Morrison), New York, Chelsea, 1956.
[XVIII]	———	"Sulla determinazione empirica di una legge di distribuzione," *Gior. Ist. Ital. Attuari*, Vol. IV-1 (1933).
[XIX]	von Kries (J.)	*Die Prinzipien der Wahrscheinlichkeitsrechnung*, Freiburg i.b., 1886.
[XX]	Lévy (P.)	*Calcul des probabilités*, Paris, Gauthier-Villars, 1925.
[XXI]	Lomnicki (A.)	"Nouveaux fondements de la théorie des probabilités," *Fundamenta Mathematica*, Vol. 4 (1923).
[XXII]	Lukasiewicz (J.)	"Philosophische Bemerkungen zu mehrwertigen Systemen des Aussagenkalküls," *Comptes Rendus Soc. Sci. Lett. Varsovie*, Classe III, Vol. XXIII (1930).
[XXIII]	Mazurkiewicz (S.)	"Zur Axiomatik der Wahrscheinlichkeitsrechnung," *C. R. Soc. Sci. Lett. Varsovie*, Classe III, Vol. XXV (1932).
[XXIV]	Medolaghi (P.)	"La logica mathematica e il calcolo delle probabilità," *Bolletino Assoc. Ital. Attuari*, **18** (1907).
[XXV]	von Mises (R.)	*Probability, Statistics and Truth*, 2nd revised English edition (prepared by H. Geringer), New York, Macmillan, 1957.

[XXVI] ———— "Théorie des probabilités: fondements et applications," *Ann. Inst. H. Poincaré,* Vol. III, fasc. 2 (1932).

[XXVII] Pareto (V.) *Manuel d'économie politique,* Paris, Giard et Brière, 1909.

[XXVIII] Poincaré (H.) *Science and Hypothesis* (translated by W. J. Greenstreet), Chapter: The Calculus of Probability, New York, Dover, 1952.

[XXIX] Reichenbach (H.) *The Theory of Probability* (English edition translated by Ernest Hutten and Maria Reichenbach), Berkeley, 1949.

[XXX] ———— "Die Induktion als Methode der wissenschaftliche Erkenntnis," Vol. IV of *Actes du Congrès Int. de Phil. Scient.* (1935), Act Scient. Indust. #391, Hermann, Paris, 1936.

[XXXI] ———— "Wahrscheinlichkeitslogik als form wissenschaftlichen Denkens," *Ibid.*

---

[16] "Sui passagi al limite nel calcolo della probabilità," *Rend. R. Ist. Lombardo,* Series 2, Vol. LXIII, fasc. 2–5 (1930).

[24] "A proposito dell'estensione del teorema delle probabilità totali alle classi numerabili," *ibid.*, fasc. 11–15 (1930).

[26] "Problemi determinati e indeterminati nel calcolo delle probabilità," *Rend. R. Acc. Naz. Lincei,* Series 6, Vol. XII, fasc. 9 (Nov., 1930).

[27] "Fondamenti logici del ragionamento probabilistico," *Boll. Un. Mat. Ital.,* Vol. IX-5 (Dec., 1930).

[28] "Ancora sull'estensione alle classi numerabili del teorema delle probabilità totali," *Rend. R. Ist. Lombardo,* Series 2, Vol. LXIII, fasc. 16–18 (1930).

[29] "Funzione caratteristica di un fenomeno aleatorio," *Memorie R. Acc. Naz. Lincei,* Series 6, Vol. IV, fasc. 5 (1930).

[32] "Probabilismo: Saggio critico sulla teoria delle probabilità e sul valore della scienze," *Biblioteca di Filosofia diretta da Antonio Aliotta,* Naples, Perrella (1931).

[34] "Sul significato soggettivo della probabilità," *Fund. Math.,* Vol. 17, Warsaw (1931).

[35] "Sui fondamenti logici del ragionamento probabilistico," *Atti. Soc. Ital. Progr. Scienze,* Riunione Bolzano-Trento del 1930, Vol. II, Rome (1931).

[36] "Le leggi differenziali e la rinunzia al determinismo," *Rend. Semin. Mat. R. Univ. Roma,* Series 2, Vol. VII (1931).

[38] "Probabilità fuori dagli schemi di urne," *Period. di Mat.,* Series 4, Vol. XII-1 (1932).

[40] "Funzione caratteristica di un fenomeno aleatorio," *Att. Congr. Int. Mathem., Bologna,* 1928, Vol. 6, Bologna, Zanichelli ed. (1932).

[45] "Sull'approssimazione empirica di una legge di probabilità," *Gior. Ist. Ital. Attuari,* Vol. 4-3 (1933).

[46 to 48] "Classi di numeri aleatori equivalenti. La legge dei grandi numeri nel caso dei numeri aleatori equivalenti. Sulla legge di distrubuzione dei valori in una successione di numeri aleatori equivalenti," *Rend. R. Acc. Naz. Lincei,* Series 6, Vol. XVIII, 2e sem., fasc. 3-4 (1933).

118 BRUNO DE FINETTI

[49]   "Sul concetto di probabilità," *Riv. Ital. Statist.*, Vol. 1 (1933).

[51]   "Indipendenza stocastica ed equivalenza stocastica," *Atti. Soc. Ital. Prog. Scienze,* Riunione Bari del 1933, Vol. II, Rome (1934).

[R. 2.]   Review of Reichenbach, H. [XXIX] *Zentralblatt für Mathematik und ihre Grenzegebiete,* 10-8 (1935).

[56]   "Il problema della perequazione," *Atti. Soc. Ital. Progr. Scienze,* Riunione Napoli del 1934, Vol. II, Rome (1935).

[58]   (In collaboration with M. M. Jacob). "Sull'integrale di Stieltjes-Riemann," *Gior. Ist. Ital. Attuari,* Vol. VI-4 (1935).

[62]   "La logique de la probabilité" (Communication to the Congrès Int. de Philosophie Scient., Paris, 1935), Act. Scient. Indus. #391, Hermann, 1936.

[64]   "Les probabilités nulles," *Bull. Sci. Math.*, Second Series, **60** (1936).

[65]   "Statistica e probabilità nella concezione di R. von Mises," *Suppl. Statis. Nuovi Probl.*, Vol. II-3 (1936).

[70]   "Riflessioni teoriche sulle assicurazioni elementari," Vol. 2 of *Comptes Rendus du Onzième Congrès International d'Actuaires* (Paris, June 17–24, 1936), Gauthier-Villars, Paris, 1937.

BERNARD O. KOOPMAN

The Bases of Probability
(1940)

The subject for consideration today[1] forms an aspect of a somewhat venerable branch of mathematical theory; but in essence it is part of a far older department of thought—the ancient science of logic. For it is concerned with a category of propositions of a nature marked by features neither physical nor mathematical, but by their rôle under the aspect of the reason. Their essential characteristic is their involvement of that species of relation between the knower and the known evoked by such terms as probability, likelihood, degree of certainty, as used in the parlance of intuitive thought. It is our threefold task to transcribe this concept into symbols, to formulate its principles, and to study its properties in their inner order and outward application.

As prelude to this undertaking it is necessary to set forth certain conventions of logic. Propositions are the elements of symbolic logic, but they may play the role of *contemplated propositions* (statements in quotation marks) or of *asserted propositions* (statements regarded as true throughout a given manipulation or deduction); and it is necessary to take account of this in the notation for the logical constants. We shall employ the symbols for negation ($\sim$), conjunction or logical product ($\cdot$), and disjunction or logical sum (v), and regard them as having no assertive power: they combine contemplated propositions into contemplated propositions and asserted propositions into asserted propositions *of the same logical type*. Quite other shall be our convention regarding implication ($\subset$) and equivalence ($=$): they combine contemplated propositions into asserted propositions, and shall not be used to combine asserted propositions in our present study. If a and b stand for contemplated propositions, the assertion that a is false (true) shall be written $a = 0$ ($a = 1$), and the assertion that a implies b, $a \subset b$ or $a \sim b = 0$; it is thus quite different from the contemplated proposition $\sim(a \sim b)$. Finally it is universally asserted that $a \sim a = 0, a \, v \sim a = 1$, and in fact all the laws of Boolean algebra are regarded as assertions. We shall assume their elements to be familiar, and shall accept without question the intuitive logical background implied in their manipulation and interpretation.[2]

(1) An address delivered before the New York meeting of the American Mathematical Society on February 24, 1940, by invitation of the Program Committee.

For the details of the theory here expounded, see the two publications of the present author. *The axioms and algebra of intuitive probability*, Annals of Mathematics, (2), vol. 41 (1940), pp. 269-292 (herein to be abbreviated as AAP) and *Intuitive probability and sequences* (forthcoming in the Annals of Mathematics) (abbreviation PS).

(2) For references see E. V. Huntington, *Transactions of the American Mathematical Society*, vol. 35 (1933), pp. 274-304.

The technical logician will observe that there is here involved a meta-mathematical question. But space forbids us to introduce and explain here the modern terminology of this subject, and compels us to throw a perhaps undue burden on the terms "contemplated" and "asserted proposition."

The next step towards our goal consists in defining a category of propositions which are to form the substratum upon which the structure of our theory is to be erected. We will designate by *experimental propositions* such statements of the outcome of a particular physical or biological event as may in principle be verified by the performance of a single crucial experiment. Thus "it will rain on this roof tomorrow at this hour" or "Mr. X made a mistake in his accounts last Monday" are experimental propositions, whereas "Newtonian mechanics is correct" is not: the motion of bodies can always be accounted for by assuming sufficiently complicated laws of force in Newton's equations, so its truth, while experimental in meaning, is not determined by a crucial experiment but rather by its ability to harmonize the results of many such experiments in an acceptably simple manner. Experimental propositions shall be denoted by lower case Latin letters; and inasmuch as a finite set of crucial experiments may always be regarded as constituting a single crucial experiment, finite combinations of letters by means of $(\sim \cdot \text{v})$ also denote experimental propositions. Finally, experimental propositions shall be regarded as contemplated propositions.[3]

At this point the category of propositions may be introduced which forms the subject of the theory of probability. We will take as the conceptual germ from which the whole theory springs the ordering of two events in the relation "not more probable than," a relation to be denoted by the partial ordering symbol ($<$)—which like ($\subset, =$) shall have assertive force. One could of course develop a theory of such assertions as $a < b$; but it would prove insufficient for the purposes of probability, as one needs for example to compare the probability of a assuming h true with its probability assuming h false. The definitive form of propositions sought is the following, in which a, b, h, k are experimental propositions and $h \neq 0, k \neq 0$:

a on the presumption h is no more

probable than b on the presumption k

and this is symbolized as $a/h < b/k$ (or equivalently, $b/k > a/h$). Thus from the four contemplated experimental propositions and the logical symbol

(3) In its occurrence in a/h the experimental proposition h is in a sense "temporarily asserted," i.e., a is viewed on the assumption that h is true. But we are applying the term asserted proposition only to those held as true on *both* sides of $a/h < b/k$.

$(/ \prec /)$ an asserted proposition $a/h \prec b/k$ (to be called a $(\prec)$ proposition) is formed which is to constitute the building stone of the whole theory of probability.

Before proceeding further, two difficulties must be surmounted.

Firstly it may be objected that $a/h \prec b/k$ is a proposition of such vague and subjective order that it may not merely be held by one person and rejected by another, but that one and the same individual may sometimes assent to it and at other times and in a different mood reject it. If this is so, how in the nature of things can such propositions form the subject matter of a precise mathematical science? Secondly it may be objected that the probability of a proposition a depends on a body of knowledge going far beyond the fact that h is true: It will involve propositions of higher logical types such as the laws of logic—and of probability itself; and perhaps even matters of subconscious moods, associations, artistic taste, and the like.

We believe that these two objections are answered at one stroke by ad-hering to the following convention, or rather, clarification of the use and laws of $(\prec)$ propositions. *A given individual at a given moment* may be regarded as assenting to a certain set of $(\prec)$ propositions; ignoring what he may hold at any other moment or what others may believe, that set of $(\prec)$ propositions which he holds at that given moment must have certain relations with one another which may be called *relations of consistency*. It is to their formulation and study that we conceive the present science to be devoted. So viewed, the analogue with strict logic is clear: many may disagree with me when I assert $a = 0$; but every one whose mind is constructed on normal lines will agree that if $a = 0$ then $(\sim\sim a) = 0$. Along with the first objection, this convention answers the second, for by positing a given indi-vidual at a given moment in the consideration of any set of $(\prec)$ propositions, the body of knowledge becomes fixed throughout, and so does not require explicit symbolization.

A third objection which might be voiced is against the restriction of the application of probability to experimental propositions. Why can we not compare the probability of two physical theories, for example? In answer to this we can say only that with the present restriction many grave logical difficulties are avoided and a theory is obtained which covers all the classical cases of mathematical probability and many others as well; and our limitation of the scope is a matter of practical strategy rather than of principle. It would undoubtedly be of interest to extend the present ideas to propositions of higher logical types, classes of propositions, logical systems and the like.

There is evidently no difference in intuitive meaning between a/h and ah/h (or, dually, $a \vee \sim h/h$), and we will make the notational convention that any one of these symbols be replaceable by any other. This has an interesting algebraic counterpart. Let $\mathcal{U}$ be the Boolean ring determined by all the experimental propositions considered in a given discussion.[4] To make the presumption that $h \in \mathcal{U}$ is true (equivalently, that $\sim h$ is false) is to make the presumption that any two propositions a and b of $\mathcal{U}$ for which $a \sim b \subset \sim h$ and $b \sim a \subset \sim h$ are equivalent, i.e., both true or false simultaneously. But this "identification" of all so-related pairs a, b is precisely the formation of the quotient ring $\mathcal{U}/(\sim h)$ whose elements $a/(\sim h)$ are the remainder classes with respect to the principal ideal $(\sim h)$. It is purely for convenience that we write a/h in lieu of $a/(\sim h)$. Thus if $\mathcal{A}$ is the class of all remainder classes in $\mathcal{U}$ with respect to all its principal ideals, the $(\prec)$ symbol introduces a partial ordering of the elements of $\mathcal{A}$.[5]

We are now ready to undertake our second task and lay down the axioms which govern any aggregate of $(\prec)$ propositions. It will be noted that they all have $(\subset, =)$ or $(\prec)$ propositions as hypothesis and as conclusion, and that in each case where the conclusion is a non-trivial $(\prec)$ proposition, this is true of the hypothesis as well. Finally, a tacit assumption is always made: *no denominator* = 0.

THE AXIOMS

V. AXIOM OF VERIFICATION. *If $k \subset b$, then $a/h \prec b/k$.*

I. AXIOM OF IMPLICATION. *If $a/h \prec b/k$ and $h \subset a$, then $k \subset b$.*

R. AXIOM OF REFLEXIVITY. *If $h = k$ and $ah = bk$, then $a/h \prec b/k$.*

T. AXIOM OF TRANSITIVITY. *If $a/h \prec b/k$ and $b/k \prec c/l$, then $a/h \prec c/l$.*

A. AXIOM OF ANTISYMMETRY. *If $a/h \prec b/k$, then $\sim a/h \succ \sim b/k$.*

C. AXIOMS OF COMPOSITION. *Let $0 \neq a_1 \subset b_1 \subset c_1$ and $0 \neq a_2 \subset b_2 \subset c_2$.*
    C_1. *If $a_1/b_1 \prec a_2/b_2$ and $b_1/c_1 \prec b_2/c_2$, then $a_1/c_1 \prec a_2/c_2$.*
    C_2. *If $a_1/b_1 \prec b_2/c_2$ and $b_1/c_1 \prec a_2/b_2$, then $a_1/c_1 \prec a_2/c_2$.*

D. AXIOMS OF DECOMPOSITION (QUASI-CONVERSES OF C). *Let $0 \neq a_1 \subset b_1 \subset c_1$, $0 \neq a_2 \subset b_2 \subset c_2$, and $a_1/c_1 \prec a_2/c_2$. Then if either symbol in* (i): *$(a_1/b_1, b_1/c_1)$ has the $(\succ)$ relation with either in* (ii): *$(a_2/b_2, b_2/c_2)$, then the remaining symbol in* (i) *has the $(\prec)$ relation with that in* (ii). (*Thus* D *contains four axioms.*)

P. AXIOM OF ALTERNATIVE PRESUMPTION. *Let $a/hb \prec r/s$ and $a/h \sim b \prec r/s$; then $a/h \prec r/s$.*

S. AXIOM OF SUBDIVISION. *For each positive integer n the following axiom is posited:*

(4) For Boolean rings and their ideals, see M. H. Stone, *Transactions of the American Mathematical Society*, vol. 40 (1936), pp. 37–111.

(5) One should guard against the notion that this ordering has any simple relation with the ordering of the remainder classes with respect to class inclusion.

S_n. *If* $a_1 \vee \cdots \vee a_n = a \neq 0$, $b_1 \vee \cdots \vee b_n = b \neq 0$, $a_i a_j = b_i b_j = 0$ *for all* $i \neq j$, *and lastly if*

$$a_1/a \prec a_2/a \prec \cdots \prec a_n/a,$$
$$b_1/b \succ b_2/b \succ \cdots \succ b_n/b,$$

then $a_1/a \prec b_1/b$.

The Axioms V, R, T, A are simply the transcription into the present language of facts so familiar as scarcely to require comment. The partial ordering property of ($\prec$) expresses itself by R and T; it leads to the definition of equiprobability $a/h \approx b/k$, inferior probability $a/h \prec b/k$ and incomparability $a/h \parallel b/k$ in the usual manner. The question of whether one can go further and assume that the entities a/h form the elements of a lattice with respect to ($\prec$) will naturally be raised; until now we have been unable to make any use of this idea, and if our experience is borne out we shall be in the presence of the first non-trivial example of a partially ordered set which is not a lattice.

Axiom I is in sharp contrast with the familiar circumstance that the numerical probability of an event may be unity (i.e., the same as a certain event) without that event's being certain. This is because numerical probability gives but a blurred rendering of the ultimate logical relations between probability and certainty.

As for Axiom C and its converse D, the following verbal rendering of C_1 may be given: If a_1 depends for its possibility on b_1, and likewise a_2 on b_2, and if c_1 is less likely to lead to b_1 than is c_2 to b_2, and if finally b_1 is in turn less likely to lead to a_1 than is b_2 to a_2, then c_1 is less likely to lead to a_1 than c_2 to a_2. So stated, it exhibits a sort of inner transitivity. All the other cases have a corresponding phraseology.

One might be disposed to regard Axiom P as a theorem which could be proved by arguing that since the hypothesis $a/hb \prec r/s$, $a/h \sim b \prec r/s$ tells us that a on the presumption h is not more likely than r/s both when b is true and when b is false, it must be so in all cases, i.e., $a/h \prec r/s$. Carrying this species of reasoning a little further, we could prove that such a relation as $a/h \approx \sim a/h$ is impossible: Since the assertion $(a \vee \sim a) = 1$ is always made we would conclude that the only possibilities are $a = 1$ or $\sim a = 1$ (i.e., $a = 0$); in either case $a/h \approx \sim a/h$ is impossible—hence it is never possible. In essence this is the old objection of the elementary student who voices it by saying that since an event will either happen or not happen, it is absurd to say that its probability of happening could ever be $\frac{1}{2}$. We all know how to answer him by general reference to the dependence of probability on a body of knowledge;[6]

(6) The notion that the uncertainty resides in the events themselves rather than in the mind of the individual contemplating them, the appeal to the "principle of uncertainty", etc., betrays merely a misconception both of probability and of quantum mechanics.

but we are in a position here to give the answer in a precise logical form: The fallacy lies in confusing the assertion $(a \vee \sim a) = 1$ with the assertion "$a = 1 \ or \sim a = 1$" [which might be written $(a = 1) \vee (\sim a = 1)$]. *The distinction between an asserted disjunction and a disjoined assertion is fundamental: $(u \vee v) = 1$ must never be confused with $(u = 1) \vee (v = 1)$.* The disregard of this distinction has led to more difficulties in the foundations of probability than is often imagined. It is now clear that the above proof of Axiom P is fallacious since it confuses $(b \vee \sim b) = 1$ with $(b = 1) \vee (\sim b = 1)$. As a matter of fact the same proof would have provided an infinite extension of Axiom P, an extension which leads to paradoxes.[7]

Axiom S is epitomized in the idea that if a first event is less likely of occurrence than its opposite and if a second is more likely than its opposite, then the first is less likely than the second. While we have not succeeded in simplifying the general case (beyond restricting S_n to prime values of n), we still feel confident that those more skilful than ourselves may have better success.

As a purely formal matter it may be remarked that in a system *completely* ordered by $(<)$, Axioms P and S are logical consequences of the rest.

Our second task being complete, we pass to the third, the deduction of all the useful theorems of probability from the axioms. But before proceeding it may be remarked that we have traversed the path of all mathematical disciplines: One proposes to study a subject of which one is made aware through the intuition, the senses, and such non-mathematical modes of perception. In undertaking it one introduces symbols and statements of axioms and laws in terms of these. But from this point on everything (except the interpretation) becomes purely mathematical: The symbolic abstractions become the ultimate objects of study, the axioms and laws become the postulates. Thenceforth we may discard the intuitionalistic introduction of our symbols and axioms and regard the latter as pure conventions (postulates) pertaining to the former, which are taken as "undefinables".

The further developments fall into three groups of theorems: the theorems on comparison, the theorems on numerical probability, and the theorems on statistical weight or frequency in a sequence.

In the first group we shall confine ourselves to citing the following typical ones. No comments appear necessary.

THEOREM. *If $ah \neq 0$ or h, then $0/1 < a/h < 1/1$.*

THEOREM. *If $a_1/h_1 < a_2/h_2$, $b_1/h_1 < b_2/h_2$, and $a_1 b_1 h_1 = a_2 b_2 h_2 = 0$, then $a_1 \vee b_1/h_1 < a_2 \vee b_2/h_2$.*

(7) Cf. PS. §2.

THEOREM. *If $a/hci \prec r/s$ $(i = 1, \ldots, n)$ and $hc_ic_j = 0$ for all $i \neq j$, then $a/hc \prec r/s$ where $c = c_1 \vee \cdots \vee c_n$.*

The second group starts with the introduction of the numerical probability $p(a/h) = p(a, h)$, i.e., the number between 0 and 1 forming the basis of the classical theory. This is accomplished as follows:

DEFINITION. *Any set of propositions $(u_1, \ldots, u_n)$ shall be called an n-scale when they satisfy the conditions* (i) $u_1 \vee \cdots \vee u_n = u \neq 0$; (ii) $u_iu_j = 0$ *(all $i \neq j$);* (iii) $u_i/u \approx u_j/u$ *(all i, j).*

ASSUMPTION. *Any positive integer n being given, the conceptional existence of at least one n-scale may be assumed.*

This is the only principle which need be assumed in addition to the axioms in all the further developments of the theory.[8] It is of a fundamentally different nature from the axioms and might be compared with the assumption so familiar in thermodynamics and other parts of physics of the possibility of a conceptual experiment.

THEOREM. *If $(u_1, \ldots, u_n)$ is an n-scale and $(v_1, \ldots, v_m)$ an m-scale, $u_1 \vee \cdots \vee u_\nu/u <, \approx, > v_1 \vee \cdots \vee v_\mu/v$ according as $\nu/n <, =, > \mu/m$.*

Let $t(n)$ be the maximum value of t for which $u_1 \vee \cdots \vee u_t/u < a/b$ holds ($t = 0$ corresponding to $0/u < a/b$) and $T(n)$ the minimum value of T for which $a/b < u_1 \vee \cdots \vee u_T/u$. By the previous theorem $t(n)$ and $T(n)$ are independent of the particular n-scale chosen; for fixed a/b they are always defined functions of n. We then prove the following:

THEOREM. *The limits $p_*(a/h) = \lim_{n \to \infty} t(n)/n$ and*
$$p^*(a/h) = \lim_{n \to \infty} T(n)/n$$
always exist, and $0 \leqslant p_*(a/h) \leqslant p^*(a/h) \leqslant 1$.

The numbers $p_*(a/h)$ and $p^*(a/h)$ may be called the lower and upper numerical probabilities of a/h.

DEFINITION. *If $p_*(a/h) = p^*(a/h)$, a/h is said to be appraisable and to have $p(a/h) = p_*(a/h) = p^*(a/h)$ as its numerical probability.*

All the classical theorems follow. We give merely the following two examples:

(8) In the exhibition of certain paradoxes the extension of the assumption regarding the existence of n-scales to $n = \aleph_0$ is required.

THEOREM. *Let $a \subset b \subset c$; if a/c are appraisable and if $p(b/c) \neq 0$, then a/b will be appraisable and $p(a/c) = p(a/b)p(b/c)$.*

THEOREM. *Let a/h and b/h be appraisable. Then $a \vee b/h$ will be appraisable if and only if ab/h is appraisable and it will then follow that $p(a/h) + p(b/h) = p(a \vee b/h) + p(ab/h)$.*

If now we consider the Boolean ring determined by the totality of propositions considered in a given discussion and assume that every a/h formed in it is appraisable, we are at the threshold of the classical theory. For we are in possession of an additive function obeying all the postulates required for its derivation. Its rôle is thus revealed as a theory of (unfaithful) numerical representation of relations belonging to the more far-reaching logical theory.

Before passing to the third group of results the question as to complete additivity is in order. Our axioms establish only the restricted additivity of numerical probability. The example of the infinite sequence of propositions $a_1, a_2, \ldots$ for which $a_1 \vee a_2 \vee \cdots = 1$, $a_i a_j = 0$ $(i \neq j)$ and $p(a_1/1) = p(a_2/1) = \cdots$ shows that the equation

$$1 = p(a_1 \vee a_2 \vee \cdots /1) = p(a_1/1) + p(a_2/1) + \cdots$$

is impossible. This could be interpreted either by regarding the assumptions concerning $a_1, a_2, \ldots$ as self-contradictory or regarding them to be valid and holding the view that complete additivity is not a general property, but occurs only in an important class of special cases where its validity is a consequence of the physical circumstances and not of the logical aspect of probability. We have adopted the latter position.

Anyone conversant with modern theoretical physics is aware of the fundamental rôle played therein by probability. Statistical mechanics is a familiar example; but even more significant is the case of quantum mechanics, the laws of which can not be stated except in terms of probability. Now if these sciences are to be regarded as affording objective pictures of nature, how can their laws involve in an essential manner the notion of probability if this is indeed a concept of logic—a mode of thought? The answer to this question is immediate: the "probability" of these branches of physics is a misnomer for statistical weight or frequency in a sequence. If an event E in such a theory can have two possible outcomes, "success" (labeled 1) and "failure" (labeled 0), a conceptually infinite sequence of trials under "the same conditions" furnishes an infinite sequence of zeros and ones (α): $(\alpha_1, \alpha_2, \ldots)$ $(\alpha_n = 0, 1)$. The physical assumption that $w = \lim_{n \to \infty} (\alpha_1 + \ldots + \alpha_n)/n$ exists is made and this statistical weight or frequency w is what is designated by the word "probability" of success of E. But the whole objective content of the physical laws in question involves solely the notion of frequency.

Yet the intuitive conception of probability upon which the present work is based plays an essential part in connection with frequency. Its rôle becomes manifest at that very moment when the experimental significance of w is sought—significance, that is, to a pre-named individual in terms of the only phenomena which can come within his ken. Then it is that we become aware that a link is needed between the finite sets of trials—all that we can actually observe—and the mathematical idealization of frequency.[9] Analysis reveals that the only possible link is bound to involve the intuitive idea of probability.[10] Granting then the present theory, are we enabled to solve the problem? That we are indeed able to give a complete and precise solution and to do so without assuming any further principles is the content of the third group of theorems, to which we now turn.

Let a_n be the experimental proposition "E succeeds at the nth trial" (so that $\alpha_n = 1$). Let h denote the statement of the common experimental condition at the instance of each trial. Then the following theorem is typical.

THEOREM. *Hypothesis* 1: $\lim_{n \to \infty} (\alpha_1 + \cdots + \alpha_n)/n = w$. *Hypothesis 2: For each positive integer t*

$$a_{i_1} \ldots a_{i_t}/h \approx a_{j_1} \ldots a_{j_t}/h,$$

where $(i_1, \ldots, i_t)$ is any set of t distinct positive integers and $(j_1, \ldots, j_t)$ a similar set. Conclusion: a_i/h is appraisable and $p(a_i/h) = w$.

The first idea which should enter the mind of the mathematician is that the sequence (α) can (at least when $0 < w < 1$) be reordered so as to yield a different frequency; yet apparently this must still be equal to w; is this not contradictory? The answer consists in examining the precise logical meaning of Hypothesis 1. Firstly, let $W(w, \mu, n)$ denote the assertion: *h implies that the number of true propositions in the set $(a_1, \ldots, a_n)$ is between $n(w - 1/\mu)$*

(9) This remains true even when frequency is thought of as a ratio in a finite sequence containing a larger number of trials than can come before the individual's observation. This is the difficulty which confronts any attempt to dispense with everything of the essence of intuitive probability and replace it by a theory of frequency. It is an attempt often made with the object of freeing the science of subjectivism (sic) and of retaining therein only an account of that which scientists "really observe"; by a strange irony it places the theory of probability completely out of contact with what any given human being could ever observe.

(10) Cf. PS, § 1.

and $n(w + 1/\mu)$. Then Hypothesis 1 becomes: For any given integer μ there exists an m such that for all $n \geqslant m$ assertion $W(w, \mu, n)$ is made. In logical symbols this is the familiar

$$\prod_{\mu=1}^{\infty} \sum_{m=1}^{\infty} \prod_{n=m}^{\infty} W(w, \mu, n).$$

This is all clear enough; the ambiguity appears when the logical form for assertion $W(w, \mu, n)$ is sought, for it turns out that there are many. The following is in a certain sense the weakest; it is the one for which the above theorem is proved; it is suitable for relating frequency to probability in physics:

$$W(w, \mu, n): \quad h \subset \sum_{(p,q)} a_{p_1} \cdots a_{p_t} \sim a_{q_1} \cdots \sim a_{q_f}.$$

Here the Σ calls for the disjunction of all terms where $(p_1, \ldots, p_t, q_1, \ldots, q_f)$ are all possible sets of $t + f$ distinct integers between 1 and n and where t is the least integer $\geqslant n(w - 1/\mu)$ and f the least integer $\geqslant n(1 - w - 1/\mu)$. But Hypothesis 1 with this form of $W(w, \mu, n)$ is not the one which makes it possible to reorder the sequence so as to produce the contradiction. For this purpose it is necessary to replace it by the entirely different $W'(w, \mu, n)$:

$$W'(w, \mu, n): \quad \sum_{(p,q)} (h \subset a_{p_1} \cdots a_{p_t} \sim a_{q_1} \cdots \sim a_{q_f}).$$

Thus the paradox is resolved as in earlier cases by maintaining the distinction between an assertion of a disjunction and a disjunction of assertions.

The complementary rôle of the hypotheses of this theorem is worthy of note. Hypothesis 1 exhausts the purely objective state of affairs, while Hypothesis 2 which expresses a sort of intuitive random quality, contains the assumption of a subjective nature which allows an actual living being to capture the otherwise inaccessible objective fact and relate it to his own world of possible experience by the agency of intuitive probability. Thus the theorem renders unto objective reality that that is objective, and unto the subjective intuition that which pertains thereto.

No discussion of the bases of probability would be complete at the present day which did not make reference to alleged cases of the violation of certain principles of classical probability by the phenomena of quantum mechanics. The precise form at which we have here arrived makes it particularly simple to subject every such case to minute scrutiny. While we have no time for examples here, we are publishing elsewhere a discussion which shows that it is the physical circumstances to which the laws of probability apply and never the laws themselves which are altered.[11] It would indeed be hard to imagine how it could be otherwise. For insofar as the laws of probability are laws of thought, they are prior to experimental verification in the laboratory. For how indeed can such experiments prove any statement? Firstly, when

(11) Cf. PS, § 6.

the statement is an experimental proposition, then a crucial experiment suffices: but this is evidently not the case for the axioms of probability, which are not experimental propositions. Secondly, when the statement introduces harmony and intelligibility into an ensemble of statements proved in the laboratory: but the axioms of probability appear rather in the rôle of the *criteria* of such harmony and intelligibility. To argue, finally, that the axioms repose on *subjective* experiments and hence are experimental in character is beside the point since we are considering quantum mechanics which is based on experiments on electron tubes and things of this sort which are hardly in a class with the subjective experiments by means of which we become aware of our own rational processes.

Having dwelt so long on the positive side, it behooves us to mention a fundamental limitation of these results. The theory cannot prove that the probability of heads on the toss of a coin is $\frac{1}{2}$. More generally, it is as impotent to derive a non-trivial ($<$) proposition from a set of propositions of whatever character not containing a ($<$) proposition (stated or implied) as are the laws of Newtonian mechanics to predict the position of a particle at a given time when no initial conditions are assigned. For after all, the whole theoretical structure is but the statement *in extenso* of the laws of consistency governing an aggregate of ($<$) propositions.

The question is naturally raised whether some further principle of a purely formal-logical nature can be enunciated which will establish ($<$) propositions *ab ovo*. Having searched high and low in the literature we become aware that every apparent case of such a principle either contains in some veiled form a ($<$) proposition in its hypothesis, or else leads to insurmountable paradoxes, as in the case of the principle of sufficient reason or symmetry of ignorance which has so long sullied the name of a priori intuitive probability. The quest for the first ($<$) proposition is epitomized by the attempt to devise an experiment proving the irrelevance of some external condition A in a trial of an event E. Such a statement of irrelevance is of course a ($<$) proposition. In order to reason that A is irrelevant to E on one occasion from the results of experiments performed on another, one must assume that certain other unavoidable differences between the two occasions are themselves irrelevant to the situation. The difficulty, exactly contrary to Napoleon's Guard, always retreats but does not expire.

It is in the light of experience such as this that we may well ask whether it is not a principle of epistemology itself that blocks our path; and, even as those who having sought in vain for perpetual motion ended by making a virtue of their failure, so we may hazard the view that in principle the authority for the first ($<$) proposition does not reside in any general law of probability, logic, or experimental science. And the notion presents itself that such primary and irreducible assumptions are grounded on a basis as much of the aesthetic as of the logical order.

I. J. GOOD

Subjective Probability as the Measure of a Non-Measurable Set

(1962)

1. Introduction

I should like to discuss some aspects of axiom systems for subjective and other kinds of probability. Before doing so, I shall summarize some verbal philosophy and terminology. Although the history of the subject is interesting and illuminating, I shall not have time to say much about it.

2. Definition

In order to define the sense in which I am using the expression "subjective probability" it will help to say what it is not, and this can be done by means of a brief classification of kinds of probability [16, 11, 8].

Each application of a theory of probability is made by a communication system that has apparently purposive behavior. I designate it as "you." It could also be called "org," a name recently used to mean an organism or organization. "You" may be one person, or an android, or a group of people, machines, neural circuits, telepathic fields, spirits, Martians and other beings. One point of the reference to machines is to emphasize that subjective probability need not be associated with metaphysical problems concerning mind (compare [7]).

We may distinguish between various kinds of probability in the following manner.

(i) Physical (material) probability, which most of us regard as existing irrespective of the existence of orgs. For example, the "unknown probability" that a loaded, but symmetrical-looking, die will come up 6.

(ii) Psychological probability, which is the kind of probability that can be inferred to some extent from your behavior, including your verbal communications.

(iii) Subjective probability, which is psychological probability modified by the attempt to achieve consistency, when a theory of probability is used combined with mature judgment.

(iv) Logical probability (called "credibility" in [19] for example), which is hypothetical subjective probability when you are perfectly rational, and therefore presumably infinitely large. Credibilities are usually assumed to have unique numerical values, when both the proposition whose credibility is under consideration and the "given" proposition are well defined. I must interrupt myself in order to defend the description "infinitely large."

You might be asked to calculate the logical probabilities of the Riemann, Fermat, and Goldbach conjectures. Each of these probabilities is either 0 or

135

1. It would be cheating to wait for someone else to produce the answers. Similarly, as pointed out in [17, p. x], you cannot predict the future state of society without first working out the whole of science. The same applies even if you are satisfied with the logical probabilities of future states of society. Therefore a rational being must have an infinite capacity for handling information. It must therefore be infinitely large, or at any rate much larger than is practicable for any known physical org. In other words, logical probabilities are liable to be unknown in practice. This difficulty occurs in a less acute form for subjective probability than for logical probability.

Attempts have been made [2, 10] to define logical probability numerically, in terms of a language or otherwise. Although such a program is stimulating and useful, the previous remarks seem to show that it can never be completed and that there will always remain domains where subjective probability will have to be used instead.

(In Carnap's contribution to this Congress he has shifted his position, and now defines logical probability to mean what I call numerically completely consistent subjective probability. He permits more than one consistent system of probabilities. Thus his present interpretation of logical probability is a consistent system within a "black box" in the sense of Section 3 below.)

Physical probability automatically *obeys* axioms, subjective probability *depends* on axioms, psychological probability neither obeys axioms nor depends very much on them. There is a continuous gradation, depending on the "degree of consistency" of the probability judgments with a system of axioms, from psychological probability to subjective probability, and beyond, to logical probability, if it exists. Although I cannot define "degree of consistency," it seems to me to have very important intuitive significance. The notion is indispensable.

In my opinion, every *measure* of a probability can be interpreted as a subjective probability. For example, the physical probability of a 6 with a loaded die can be estimated as equal to the subjective probability of a 6 on the next throw, after several throws. Further, if you can become aware of the value of a logical probability, you would adopt it as your subjective probability. Therefore a single set of axioms should be applicable to all kinds of probability (except psychological probability), namely the axioms of subjective probability.

Superficially, at least, there seems to be a distinction between the axiom systems that are appropriate for physical probability and those appropriate for subjective probability, in that the latter are more often expressed in terms of inequalities, i.e., comparisons between probabilities. Theories in which inequalities are taken seriously are more general than those in which each probability is assumed to be a precise number. I do not know whether physical probabilities are absolutely precise, but they are usually assumed to be, with a resulting simplification in the axioms.

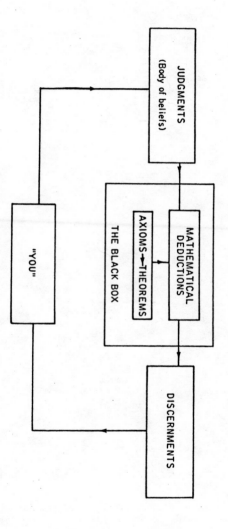

Fig. 1. The black-box flow diagram for the application of formalized scientific theories.

3. A Black-Box Description of the Application of Formalized Theories

I refer here to a "description," and not to a "theory," because I wish to avoid a discussion of the theory of the application of the black-box theory of the application of theories [4, 5, 6]. The description is in terms of the block diagram of Fig. 1 in which observations and experiments have been omitted. It consists of a closed loop in which you feed judgments into a black box and feed "discernments" out of it. These discernments are made in the black box as deductions from the judgments and axioms, and also, as a matter of expediency, from theorems deduced from the axioms alone. If no judgments are fed in, no discernments emerge. The totality of judgments at any time is called a "body of beliefs." You examine each discernment, and if it seems reasonable, you transfer it to the body of beliefs. The purpose of the deductions, in each application of the theory, is to enlarge the body of beliefs, and to detect inconsistencies in it. When these are found, you attempt to remove them by means of more mature judgment.

The particular scientific theory is determined by the axioms and the rules of application.

The rules of application refer to the method of formalizing the judgments, and of "deformalizing" the mathematical deductions. For example, in a theory of subjective probability the standard type of judgment might be a comparison of the form

$$P'(E|F) \geqq P'(G|H),$$

FIG. 1. The black-box flow diagram for the application of formalized scientific theories.

where $P'(E|F)$ is the intensity of conviction or degree of belief that you would have in E if you regarded F as certain. The P''s are not necessarily numerical, and what is meaningful is not a P' by itself, but a comparison of intensities of conviction of the above type. These judgments are plugged into the black box by simply erasing the two dashes. Likewise, discernments can be obtained by taking an output inequality, $P(E|F) \geqq P(G|H)$, and putting dashes on it. The P's are assumed to be numbers, even if you can never discover their values at all precisely. This is the reason for the expression "black box." The black box may be entirely outside you, and used like a tame mathematician, or it may be partially or entirely inside you, but in any case you do not know the P's precisely.

Following Keynes and Koopman, I assume that the P''s are only partially ordered.

Apart from the axioms and rules, there are in practice many informal suggestions that you make use of, such as the need to throw away evidence judged to be unimportant, in order to simplify the analysis, and yet to avoid special selection of the evidence. (In spite of all the dangers, objectivistic methods in statistics invariably ignore evidence in order to achieve objectivity. In each application a subjective judgment is required in order to justify this device. Compare the idea of a "Statistician's Stooge" in [9].) But in this paper I am more concerned with axioms and rules of application than with "suggestions."

De luxe black boxes are available with extra peripheral equipment, so that additional types of judgment and discernment can be used, such as direct judgments of odds, log-odds, "weights of evidence," numerical probabilities, judgments of approximate normality, and (for a theory of rational behavior) comparisons of utilities and of expected utilities [5 or 6]. (There are numerous aids to such judgments, even including black-box theorems, such as the central limit theorem, and a knowledge of the judgments of other orgs. All such aids come in the category of "suggestions.") But, as we shall see shortly, for the simplest kind of black box, a certain kind of output must not be available.

4. Axiom Systems for Subjective Probability

See, for example, [18, 3, 13, 14, 4, 20] and, for similar systems for logical probability, [12, 10]. The axioms of subjective probability can be expressed in terms of either

(i) comparisons between probabilities, or preferences between acts, or
(ii) numerical probabilities.

Koopman's system [13, 14] was concerned with comparisons between probabilities, without reference to utilities or acts. Although it is complicated it is convincing when you think hard. From his axioms he deduced numerical ones for what he called *upper* and *lower* probabilities, P^* and P_*. We may define $P_*(E|F)$ and $P^*(E|F)$ as the least upper bound and greatest lower bound of numbers, x, for which you can judge or discern that $P'(E|F) > x$ or $< x$. Here $P'(E|F)$ is not a number, although x is. The interpretation of the inequality $P'(E|F) > x$ is as follows. For each integer x, perfect packs of n cards, perfectly well shuffled are imagined so that for each rational number, $x = m/n(m < n)$, there exist propositions, G and H, for which $P(G|H)$ would usually be said to be equal to x. The inequality is then interpreted as $P'(E|F) > P'(G|H)$.

Note that $P_*(E|F)$ and $P^*(E|F)$ depend on the whole body of beliefs. Also note that $P_*(E|F)$ is not the least upper bound of all numbers, x, for which you can *consistently* state that $P'(E|F) > x$: to assume this interpretation for more than one probability would be liable to lead to an *inconsistency.*

If $P_* = P^*$, then each is called P, and the usual axioms of probability are included in those for upper and lower probability. The analogy with inner and outer measure is obvious. But the axioms for upper and lower probability do not all follow from the theory of outer and inner measure. It is a little misleading to say that probability theory is a branch of measure theory.

In order to avoid the complications of Koopman's approach, I have in the past adopted another one, less rigorous, but simpler. I was concerned with describing how subjective probability should be used in as simple terms as possible more than with exact formal justification. (I gave an informal justification which I found convincing myself.) This approach consisted in assuming that a probability *inside the black box* was numerical and precise. This assumption enables one to use a simple set of axioms such as the following set (axioms C).

C1. $P(E|F)$ *is a real number.* (Here and later, the "given" proposition is assumed not to be self-contradictory.)

C2. $0 \leq P(E|F) \leq 1$.

DEFINITION. *If* $P(E|F) = 0$ (or 1), *then we say that E is "almost impossible" (or "almost certain") given F.*

C3.   *If E.F is almost impossible given H, then*

$$P(E \vee F |H) = P(E|H) + P(F|H) \text{ (addition axiom).}$$

C4.   *If H logically implies E, then E is almost certain given H (but not conversely).*

C5. *If H·E and H·F are not self-contradictory and H·E implies F and H·F implies E, then*

$$P(E|H) = P(F|H) \text{ (axiom of equivalence)}.$$

C6. $P(E·F|H) = P(E|H) · P(F|E · H)$ (product axiom).

C7. (Optional.) *If $E_i·E_j$ is almost impossible given H ($i < j$; i, j = 1, 2, 3, ⋯ ad inf.), then*

$$P(E_1 \vee E_2 \vee \cdots | H) = \sum_i P(E_i | H) \text{ (complete additivity)}.$$

(The above axioms are not quite the same as axiom sets *A* and *B* of [4].)

C8. (The Keynes-Russell form of the principle of cogent reason. Optional. See [19, p. 397], [4, p. 37].) *Let ϕ and ψ be propositional functions. Then*

$$P(\phi(a)| \psi(a)) = P(\phi(b)| \psi(b)).$$

I describe this axiom as "optional" because I think that in all those circumstances in which it is judged to be (approximately) applicable, the judgment will come to the same thing as that of the equation itself, with dashes on.

It follows from axiom C1, that $P(E|F) <, >,$ or $= P(G|H)$, but we do not want to deduce that $P'(E|F)$ and $P'(G|H)$ are comparable. There is therefore an artificial restriction on what peripheral equipment is available with *de luxe* black boxes. This artificiality is the price to be paid for making the axioms as simple as possible. It can be removed by formulating the axioms in terms of upper and lower probabilities. To use axioms C is like saying of a non-measurable set that it really has an unknowable ("metamathematical") measure lying somewhere between its inner and outer measures. And as a matter of fact there is something to be said for this paradoxical-sounding idea. If you will bear with me for a moment I shall illustrate this in a non-rigorous manner.

Suppose *A* and *B* are two non-intersecting and non-measurable sets. Write *m* for the unknowable measure of a non-measurable set, and assume that

$$m(A + B) = m(A) + m(B).$$

Then

$$m(A + B) \leq m^*(A) + m^*(B), \ m(A + B) \geq m_*(A) + m_*(B).$$

Therefore (for elucidation, compare the following probability argument)

$$m^*(A + B) \leq m^*(A) + m^*(B), \quad m_*(A + B) \geq m_*(A) + m_*(B),$$

and these inequalities are true. Similarly,

$$m(A) = m(A + B) - m(B) \leq m^*(A + B) - m_*(B).$$

Therefore

$$m^*(A) \leq m^*(A + B) - m_*(B),$$

which is also true.

The same metamathematical procedure can be used, more rigorously, in order to derive without difficulty, from axioms C together with the rules of application, a system of axioms for upper and lower probability. These are the axioms D listed below. As an example, I shall prove axiom D6(iii). We have

$$P(E \cdot F | H) = P(E | H) \cdot P(F | E \cdot H).$$

Therefore, if $P(F | E \cdot H) \neq 0$,

$$P(E \cdot F | H) / P(F | E \cdot H) = P(E | H).$$

But $P^*(E \cdot F | H) \geq P(E \cdot F | H)$, since, in this system, $P^*(E \cdot F | H)$ is defined as the greatest lower bound of numbers, x, for which it can be discerned that $x > P(E \cdot F | H)$. Similarly,

$$P_*(F | E \cdot H) \leq P(F | E \cdot H).$$

Therefore

$$P^*(E \cdot F | H) / P_*(F | E \cdot H) \geq P(E | H).$$

Therefore

$$P^*(E \cdot F | H) / P_*(F | E \cdot H) \geq P^*(E | H).$$

Q.E.D.

The main rule of application is now that the judgment or discernment $P'(E | F) > P'(G | H)$ corresponds to the black-box inequality $P_*(E | F) >$

$P^*(G|H)$. Koopman derived most, but not all, of axioms D1-D6 from his non-numerical ones, together with an assumption that can be informally described as saying that perfect packs of cards can be imagined to exist. His derived axioms for upper and lower probability do not include axiom D6(iii) and (iv). (D7 and D9 were excluded since he explicitly avoided complete additivity.) I have not yet been able to decide whether it is necessary to add something to his non-numerical axioms in order to be able to derive D6(iii) and (iv). Whether or not it turns out to be necessary, we may say that the present metamathematical approach has justified itself, since it leads very easily to a more complete set of axioms for upper and lower probability than were reached by Koopman with some difficulty.

The axioms D will now be listed. I have not proved that the list is complete, i.e., that further independent deductions cannot be made from axioms C.

D1. *$P_*(E|F)$ and $P^*(E|F)$ are real numbers.* (Here and later the given proposition is assumed not to be self-contradictory.)

D2. $0 \leq P_*(E|F) \leq P^*(E|F) \leq 1$.

DEFINITION. *If $P^* = P_*$, each is called P. The previous definitions of "almost certain" and "almost impossible" can then be expressed as $P_* = 1$ and $P^* = 0$.*

D3. *If $E \cdot F$ is almost impossible given H, then* (addition axiom)

$$P_*(E|H) + P_*(F|H) \leq P_*(E \vee F|H) \leq P_*(E|H) + P^*(F|H)$$
$$\leq P^*(E \vee F|H) \leq P^*(E|H) + P^*(F|H).$$

D4. *If H logically implies E, then E is almost certain given H but not conversely.*

D5. *If $H \cdot E$ implies F, and $H \cdot F$ implies E, then* (axiom of equivalence)

$$P_*(E|H) = P_*(F|H), \; P^*(E|H) = P^*(F|H).$$

D6. (Product axiom.)

(i) $\qquad\qquad P_*(E \cdot F|H) \geq P_*(E|H) \cdot P_*(F|E \cdot H);$

(ii) $\qquad\qquad P^*(E \cdot F|H) \geq P_*(E|H) \cdot P^*(F|E \cdot H);$

(iii) $\qquad\qquad P^*(E \cdot F|H) \geq P^*(E|H) \cdot P_*(F|E \cdot H);$

(iv) $\qquad\qquad P_*(E \cdot F|H) \leq P_*(E|H) \cdot P^*(F|E \cdot H);$

(v) $\qquad\qquad P_*(E \cdot F|H) \leq P^*(E|H) \cdot P_*(F|E \cdot H);$

(vi) $\qquad\qquad P^*(E \cdot F|H) \leq P^*(E|H) \cdot P^*(F|E \cdot H).$

D7. (Complete super- and sub-additivity. Optional.)

If $E_i \cdot E_j$ is almost impossible given H ($i < j$; $i, j = 1, 2, 3, \cdots$ ad inf.), then

(i) $P_*(E_1|H) + P_*(E_2|H) + \cdots \leq P_*(E_1 \vee E_2 \vee \cdots |H)$;

(ii) $P^*(E_1|H) + P^*(E_2|H) + \cdots \geq P^*(E_1 \vee E_2 \vee \cdots |H)$.

D8. (Cogent reason. Optional.) *Let ϕ and ψ be propositional functions. Then*

$$P_*(\phi(a)|\psi(a)) = P_*(\phi(b)|\psi(b)), \quad P^*(\phi(a)|\psi(a)) = P^*(\phi(b)|\psi(b)).$$

D9. (Complete super- and sub-multiplicativity. Optional.) *For any (enumerable) sequence of propositions, $E_1, E_2, \ldots$.*

(i) $P_*(E_1|H) P_*(E_2|E_1 \cdot H) P_*(E_3|E_1 \cdot E_2 \cdot H)$
      $\cdots \leq P_*(E_1 \cdot E_2 \cdots |H)$;

(ii) $P^*(E_1|H) P^*(E_2|E_1 \cdot H) P^*(E_3|E_1 \cdot E_2 \cdot H)$
      $\cdots \geq P^*(E_1 \cdot E_2 \cdots |H)$.

The corresponding result in the C system is a *theorem.* (See Appendix.)
I have not been able to prove that

$$P^*(E \vee F) + P^*(E \cdot F) \leq P^*(E) + P^*(F),$$

even though the corresponding property is true of Lebesgue outer measure, and I suspect that it does not follow by the above methods. It would be possible to prove it (compare [1, p. 14]) provided that we assumed:

D10. (Optional.) *Given any proposition, E, and a positive number, ε, there exists a proposition, G, which is implied by E, and has a precise probability $P(G) < P^*(E) + \varepsilon$.* (This axiom may be made conditional on another proposition, H, in an obvious manner.)

I cannot say that D10 has much intuitive appeal, although the corresponding assertion is true in the theory of measure.

It seems that the theory of probability is not quite a branch of the theory of measure, but each can learn something from the other.

Incidentally, random variables can be regarded as *isomorphic* with arbitrary functions, not necessarily measurable. I understand that this thesis is supported by de Finetti. Also upper and lower expectations can be defined by means of upper and lower integrals in the sense of Stone [21].

5. Higher Types of Probability

A familiar objection to precise numerical subjective probability is the sarcastic request for an estimate correct to twenty decimal places, say for the probability that the Republicans will win the election. One reason for using upper and lower probabilities is to meet this objection. The objection is however raised, more harmlessly, against the precision of the upper and lower probabilities. In [5], and in lectures at Princeton and Chicago in 1955, I attempted to cope with this difficulty by reference to probabilities of "higher type." When we estimate that a probability $P'(E|H)$ lies between 0.2 and 0.8 we may feel that 0.5 is *more likely to be rational* than 0.2 or 0.8. The probability involved in this expression "more likely" is of "type II." I maintain that we can have a subjective probability distribution concerning the estimate that a perfectly rational org would make for $P'(E|H)$, a subjective probability distribution for a credibility. *If* this probability distribution were sharp, then it could be used in order to calculate the expected credibility precisely, and this expectation should then be taken as our subjective probability of E given H. But the type II distribution is not sharp; it is expressible only in terms of inequalities. These inequalities themselves have fuzziness, in fact the fuzziness obviously increases as we proceed to higher types of probability, but it becomes of less practical importance.

It seems to me that type II probability is decidedly useful as an unofficial aid to the formulation of judgments of upper and lower probability of type I. I would not myself advocate even the unofficial use of type III probability for most practical purposes, but the notion of an infinite sequence of types of probability does have the philosophical use of providing a rationale for the lack of precision of upper and lower probabilities.

Appendix. Continuity.

(See the remark following D9.) There is a well-known strong analogy between the calculus of sets of points and the calculus of propositions. In this analogy "E is contained in F" becomes "E implies F"; $E + F$ becomes $E \lor F$; $E - F$ becomes $E \cdot \overline{F}$; $E \cap F$ becomes $E \cdot F$; "E is empty" becomes "E is impossible"; "all sets are contained in E" becomes "E is certain"; $E_n \nearrow$ becomes $E_1 \supset E_2 \supset E_3 \supset \cdots$; $E_n \searrow$ becomes $\cdots E_3 \supset E_2 \supset E_1$.

Accordingly we can define, for an infinite sequence of propositions, $\{E_n\}$,

$$\lim \sup E_n = (E_1 \lor E_2 \lor \cdots) \cdot (E_2 \lor E_3 \lor \cdots) \cdot (E_3 \lor E_4 \lor \cdots) \cdots,$$
$$\lim \inf E_n = (E_1 \cdot E_2 \cdots) \lor (E_2 \cdot E_3 \cdots) \lor (E_3 \cdot E_4 \cdots) \lor \cdots.$$

If these are equal, each is called $\lim E_n$. The limit of a monotonic increasing (or decreasing) sequence of propositions is

$$E_1 \lor E_2 \lor \cdots \text{ (or } E_1 \cdot E_2 \cdots \text{)}.$$

The other definitions and arguments given, for example in [15, pp. 84-85] can be at once adapted to propositions, and we see that complete additivity is equivalent to continuity, i.e., lim $P(E_n) = P$ (lim E_n) if $\{E_n\}$ is a monotonic sequence of propositions. It can be proved that, for example,

$$P(E_1 \cdot E_2 \cdots) = P(E_1) \cdot P(E_2 | E_1) \cdot P(E_3 | E_1 \cdot E_2) \cdots ,$$

i.e., we have "complete multiplicativity." The axiom D9 is derived from this theorem by means of the metamathematical argument of Section 4.

The analogy between propositions and sets of points is imperfect. For the analogy of a point itself should be a logically possible proposition, E, that is not implied by any other distinct proposition that is logically possible. It is not easy to think of any such proposition E, unless the class of propositions has been suitably restricted. Fortunately the notion of a point is inessential for the above analysis: the algebra of sets of points makes little use of the notion of a point.

REFERENCES

[1] Burkill, J. C. *The Lebesgue Integral,* Cambridge, University Press, 1951.

[2] Carnap, R. *Logical Foundations of Probability.* Chicago, University of Chicago Press, 1950.

[3] Finetti, B. de. La prévision: ses lois logiques, ses sources subjectives. *Annales de l'Inst. Henri Poincaré,* Vol. 7 (1937), pp. 1-68.

[4] Good, I. J. *Probability and the Weighing of Evidence,* London, Griffin; New York, Hafner, 1950.

[5] Good, I. J. Rational decisions. *Journal of the Royal Statistical Society, Series B,* Vol. 13 (1952), pp. 107-114.

[6] Good, I. J. Chapter 3 of *Uncertainty and Business Decisions,* 2nd ed., Liverpool, University Press, 1957. (Ed. by Carter, Meredith and Shackle.)

[7] Good, I. J. Could a machine make probability judgments? *Computers and Automation,* Vol. 8 (1959), pp. 14-16 and 24-26.

[8] Good, I. J. Kinds of probability. *Science,* Vol. 129 (1959), pp. 443-447.

[9] Good, I. J. The paradox of confirmation. *British Journal of the Philosophy of Science,* Vol. 11 (1960), pp. 145-149; Vol. 12 (1961), pp. 63-64.

[10] Jeffreys, H. *Theory of Probability.* Oxford, University Press, 1939.

[11] Kemble, E. C. The probability concept. *Philosophy of Science,* Vol. 8 (1941), pp. 204-232.

[12] Keynes, J. M. *A Treatise on Probability.* London, Macmillan, 1921.

[13] Koopman, B. O. The axioms and algebra of intuitive probability. *Annals of Mathematics,* Vol. 41 (1940), pp. 269-292.

[14] Koopman, B. O. The bases of probability. *Bulletin of the American Mathematical Society,* Vol. 46 (1940), pp. 763-774.

[15] Loève, M. *Probability Theory,* Toronto, New York, London, van Nostrand, 1955.

[16] Poisson, S. D. *Recherches sur la Probabilité des Jugements.* Paris, Bachelier, 1837.

[17] Popper, K. R. *The Poverty of Historicism.* London, Routledge and Kegan Paul, 1957.

[18] Ramsey, F. P. Chapters 7 and 8 of *The Foundations of Mathematics,* London, Routledge and Kegan Paul, 1931.
[19] Russell, B. *Human Knowledge, its Scope and Limits,* London, George Allen, 1948.
[20] Savage, L. J. *The Foundations of Statistics,* New York, Wiley; London, Chapman and Hall, 1954.
[21] Stone, M. H. Notes on integration, I, II, III, IV. *Proceedings of the National Academy of Sciences,* Vols. 34 and 35 (1948).

L. J. SAVAGE

The Elicitation of Personal Probabilities and Expectations

(1971)

© *Journal of the American Statistical Association,* December 1971, Volume 66, Number 336, Theory and Methods Section. Reprinted by permission of the editors.

Proper scoring rules, i.e., devices of a certain class for eliciting a person's probabilities and other expectations, are studied, mainly theoretically but with some speculations about application. The relation of proper scoring rules to other economic devices and to the foundations of the personalistic theory of probability is brought out. The implications of various restrictions, especially symmetry restrictions, on scoring rules is explored, usually with a minimum of regularity hypothesis.

1. INTRODUCTION*

1.1 Preface

This article is about a class of devices by means of which an idealized *homo economicus*—and therefore, with some approximation, a real person—can be induced to reveal his opinions as expressed by the probabilities that he associates with events or, more generally, has personal expectations of random quantities. My emphasis here is theoretical, though some experimental considerations will be mentioned. The empirical importance of such studies in many areas is now recognized. It was emphasized for the area of economics in an address by Trygve Haavelmo [28, p. 357]:

> I think most of us feel that if we could use *explicitly* such variables as, e.g., what people *think* prices or incomes are going to be, or variables expressing what people *think* the effects of their actions are going to be, we would be able to establish relations that could be more accurate and have more explanatory value. But because the statistics on such variables are not very far developed, we do not take the formulation of theories in terms of these variables seriously enough. It is my belief that if we can develop more explicit and a priori convincing economic models in terms of these variables, which are realities in the minds of people even if they are not in the current statistical yearbooks, then

* Leonard J. Savage is Eugene Higgins Professor of Statistics and chairman, Department of Statistics, Yale University, New Haven, Conn. 06520. This article was prepared in connection with research supported by the Army, Navy, Air Force and NASA under contracts administered by the Office of Naval Research. Reproduction in whole or in part is permitted for any purpose of the United States Government. Part of the work was supported by the Michigan Institute of Science and Technology. The author wishes to thank Ward Edwards, Allan Murphy, Richard Savage, Carl-Axel Stael von Holstein, Robert Winkler and an excellent referee for their help in improving the article. He is especially grateful to Bruno de Finetti who provided the impetus for the article.
EDITOR'S NOTE: Professor Savage died suddenly on November 1, 1971, while this article was in press. A memorial article on his work will appear in a forthcoming issue of one of the ASA publications.

ways and means can and will eventually be found to obtain actual measurements of such data.

A special instance of the central general principle of this article was recognized long ago by Brier [5], the general principle itself was briefly but colorfully announced by McCarthy [37], and a considerable literature pertaining to it has grown up, some of which will be cited in context and most of which can be found through the references cited, especially the recent and extensive [52] and others that I call "key references."

Bruno de Finetti and I began to write the present article in the spring of 1960, not yet aware of our predecessors and contemporaries. The impetus was de Finetti's, for he had brought us to rediscover McCarthy's [37] insight about convex functions. We expected to make short work of our "little note," but it grew rapidly in many directions and became inordinately delayed. Now we find that the material in the present article is largely mine and that de Finetti has published on diverse aspects of the same subject elsewhere [12, 13, 14, 17]. De Finetti has therefore withdrawn himself from our joint authorship and encouraged me to publish this article alone, though it owes so much to him at every stage, including the final draft.

The article is written for a diverse audience. Consequently, some will find parts of it mathematically too technical, and others will find parts too elementary. If each skips what puzzles or bores him he will, I hope, find the rest reasonably complete for him.

1.2 Summary

The bare essentials of the economic theory of personal probability and expectation are introduced (Section 2). Various difficulties in principle that beset the evaluation of preferences such as those that determine the price at which a man is just willing to sell his car or the probability for him that the car will shortly need a new muffler are discussed (Section 3).

A probability is a price, in a manner of speaking. More accurately, it is a marginal rate of substitution. Such rates can be elicited by a general mode of behavioral interrogation rather like one that has been proposed for ordinary prices (Section 4). These methods admit mathematically special cases that seem to be of particular interest (Section 5).

The rate-eliciting methods, both general and special, are examined with reference to probabilities and expectations (Section 6). The methods of eliciting personal probabilities thus arrived at can also be approached through the ideas of statistical decision theory (Section 7), and this decision-theoretic method provides a relatively new basis due to de Finetti for defining personal probability (Section 8).

The methods are generalized to simultaneous elicitation of several rates, probabilities, or expections (Section 9). This vectorial discussion concludes with a prominent special case in which the subject's income depends only on that event among disjoint and exhaustive possibilities that actually occurs.

Aspects of possible applications are discussed (Section 10). These include domains of possible application, impediments to application, and criteria for choice among the methods.

2. RATES OF SUBSTITUTION AND PROBABILITY AND EXPECTATION

A man who owns some of a commodity, say wheat, will, under suitable circumstances, be almost indifferent both to buying and selling modest quantities of it at a certain price per bushel, his (marginal) rate of substitution of cash for wheat.

Money payable subject to a contingency, such as the accidental burning of a house or the outcome of a race, can be regarded as a commodity [4]. Such commodities are explicitly dealt in by insurance companies and bookmakers, and we encounter them implicitly wherever we make decisions in the face of uncertainty. There is reason to postulate that an ideally coherent person has a rate of substitution $P(A)$ for money contingent on the event A. When q is not too large, he is indifferent to buying or selling q dollars contingent on A for $qP(A)$ dollars outright, and $P(A)$ is defined as the probability of A for the person [9]. Though the relationship can, because of the nonlinearity of the utility of money, be expected to hold only in the limit for infinitesimal values of q, I shall usually write as though it were exact. This limitation is serious, though a technique for creating utility-free currency and other ways to avoid the effects of nonlinear utility will be mentioned in the next section.

Let U denote the logically certain event and A and B any two logically incompatible events. For an economically coherent person,

$$P(U) = 1, \ P(A) \geqslant 0, \ P(A \text{ or } B) = P(A) + P(B), \tag{2.1}$$

as shown in [9]. Thus, such a person's P is a (finitely additive) probability measure.

If a finite sequence $\{ A_s \}$ of events is a partition (that is, if every pair of them is incompatible but one or another of them obtains), then

$$\Sigma P(A_s) = 1, \tag{2.2}$$

as follows from (2.1).

To orient the reader in the critical literature on personal probability, the commentary and bibliography of [47] might be useful. Excellent early papers are republished, in English, in [31]. Some more recent ones are [51, 18, 16, 17]. An extensive bibliography with special reference to experimental aspects is [19].

More general than the notion of the probability of an event is that of the expectation of a random quantity. Let $\{A_s\}$ be a partition and $\{v_s\}$ a corresponding sequence of numbers; the two together can be regarded as the random quantity V that takes the value v_s if A_s obtains. If a person's probability of A_s is p_s, what ought he pay for the simultaneous offer of qV dollars, that is, for qv_1 dollars in case A_1 obtains, qv_2 dollars in case A_2 obtains, etc.? Viewing qV as a commodity bundle consisting of qv_1 units of one commodity, qv_2 unit of another, etc., the answer (always for moderate q) is clearly

$$qE(V) = q \sum p_s v_s,\tag{2.3}$$

or q times the expectation of V.

A probability $P(A)$ is itself plainly an expectation, namely $E(I_A)$, where I_A is the random quantity that is 1 if A obtains and 0 if A does not obtain. The event A and its associated random quantity I_A can, to considerable advantage, be rigorously regarded as two aspects of one object, simply denoted by A [15]. Thus it is meaningful and convenient to write $P(A) = E(A)$.

A random quantity V need not be defined in terms of a partition nor need it have only a finite number of possible values; it may be simply any (ordinarily) unknown, empirically determinable number. For the unknown payment qV, the person is presumably willing to exchange some definite payment $qE(V)$. The rate of substitution $E(V)$ is the person's expectation of the unknown, or random, quantity V. (In practice, V would be bounded; mathematically, it can be useful to consider some unbounded V and to exclude others.)

3. SOME DIFFICULTIES IN THE EXPERIMENTAL ELICITATION OF PREFERENCES

The difficulties mentioned in this section are mainly tangential to the present article. Some of them have little or no bearing on the particular situation to be studied; others are here treated as secondary for the time being. The first type are mentioned only to emphasize the advantage of a certain method of eliciting prices and rates of substitution, including probabilities and expectations; and the others are mentioned to warn of their existence.

Many conceptual experiments on preference will presumably remain conceptual only, some because their financial cost is prohibitive and others because they imply immoral or impractical interference with people's lives. For example, experiments to determine directly what risk of pauperhood a person will take to avoid spending a hundred dollars or to gain a million dollars are literally fantastic. This does not of course preclude learning something indirectly: from the behavior of buyers of insurance and of gamblers and speculators [23, 24]; or by asking subjects to introspect about hypothetical choice; or by observing changes of economic behavior that follow changes in the policies of governments, firms, and other institutions; or in other ways.

Statistical problems and difficulties arise in empirical studies of preference, as they do in all empirical studies. A more special, but perhaps related difficulty, is that all subjects report, or otherwise reveal, that they do not know their own preferences; they experience wavering and indecision that cannot be identified with mere indifference. See, for example, [47, pp. 21, 59, 168-9; 18, Section 26; 45; and 22 under "Indifference, Intransitive" in the Index].

Another difficulty peculiar to experiments on preference is that once an experimenter satisfies one preference of a subject, he may quite drastically change the subject's pattern of preferences. The thirsty man who now prefers water to wine and wine to whiskey might fail to reveal this in an experiment in which he is first offered his choice between wine and water and then, his thirst quenched with water, is offered his choice between wine and whiskey. A related phenomenon is illustrated by the subject who accepts whiskey (perhaps for later consumption) in preference to wine because he suspects that the experimenter will shortly be offering him an opportunity to obtain water. Such difficulties arise in any attempt to study the gambling (and insuring) preferences of a subject. They are particularly evident and important in experiments to determine the demand price or the offer price of a subject for a specific object or service.

One interesting device for coping with such difficulties was pointed out to me by W. Allen Wallis in 1949 or 1950 and was independently exploited by Allai [2] in experiments conducted in 1952. Let the experimenter put the subject successively in several hypothetical choice situations always with the understanding that, when all the subject's choices have been expressed, one of them will actually be implemented by a chance device that, plain to the subject, has no direct connection with the situations of interest. For instance, a subject asked to rank half a dozen tickets to plays, concerts, and athletic events in the order of his preference will, insofar as he is rational, do so sincerely if he understands that the tickets are to be thoroughly shuffled and that he will be given that one of the top two tickets for which he has indicated preference.

The following application of the same general principle was introduced by Marschak [33, 34]. The experimenter makes a sealed bid for an object in the subject's possession and the subject, before seeing this bid, puts an "asking price" on the object. The sale takes place at the bid price, if and only if the bid is at least as high as the asking price. It is clearly to the advantage of the subject to name his actual offer price, at least if he is not in a position to exclude the possibility of bids in some neighborhood of this price; and even in that exceptional case, there is no advantage in naming any other price. (The procedure can evidently be so modified that the subject is the buyer rather than the seller.)

Actual subjects are of course sometimes blind to their own clear advantage [34; 35, p. 47], failing, for example, to understand that they can only deprive themselves by asking too much. It is no final criticism of such a method to say that subjects do not automatically and instinctively understand it or that, understanding, they have psychological difficulty in doing the rational thing. Such facts do underline the need for education and training prior to, and even during, the application of elicitation devices. Incidentally, such education promises to be of great general benefit to the subject and deserves wide promulgation on its own account.

Marschak's method does not depend on the hypothesis that the utility of money is linear, but throughout most of this article, this hypothesis is relied upon. The hypothesis is presumably a valid approximation if only small transactions are involved. Still better, there would be no approximation at all if payments were made in utiles rather than in money, which may sometimes be roughly feasible.

A certain scheme for effecting payments in utiles is vividly suggested by Smith [50, Section 13]. If, to paraphrase Smith, the currency used consists of tickets in a one-prize lottery, which the subject is known to regard as fair and independent of all uncertainties that are of direct interest for the experiment, then the subject's utility for this currency is linear, though the transactions may be very substantial. Of course for this scheme to be valid, the utility of the prize to the subject must not depend on the outcome of events that are of interest. For example, a lottery in which the prize is a diamond would obviously not provide a valid utility-free currency for exploring a subject's opinions about the future of the diamond market.

In principle, the utility function of a subject for ordinary money can be determined by certain elicitation experiments that give operational meaning to utility. With this function known, the experimenter would know the utility worth to the subject of any proposed payment (positive or negative) and could arrange to make payments in utiles. Practical limitations on this scheme and on the one in the preceding paragraph are severe, but the ideas are at least sufficiently suggestive to be worth bringing out [57, 60].

The linear approximation seems adequate for many applications. Where it is not quite adequate, much can be said for taking as the next approximation a function for the utility of money of the form

$$(1 - c^{-\lambda x})/\lambda. \tag{3.1}$$

(Any function of the form $b - ac^{-\lambda x}$, with a and λ of like sign, would of course amount to the same thing; the choice of the constants a and b made in (3.1) emphasize that the proper interpretation of $\lambda = 0$ is linear utility, because, for fixed x and small λ, (3.1) is approximately $[1 - (1 - \lambda x)]/\lambda = x$.) For almost all applications λ would be positive, which is why a minus sign was introduced in the exponent of (3.1). A person who is just willing to toss a fair coin for $10.00 if he receives a side payment of $0.10 exhibits a λ of about .0020 per dollar. To show that a person's utility for money is adequately described by (3.1) in some range of practical interest and to determine the value applicable to him may in some contexts provide a practical basis for paying him in utiles.

Taken literally, the exponential utilities described what I would call a perfect miser, a person who does not necessarily rank lotteries, and the like, simply by their expected cash value but whose rankings do not fluctuate with his own wealth at the moment. Within sufficiently narrow limits, any person's utilities can be expected to be practically linear. When these limits are somewhat exceeded so that nonlinearity must be taken into account, there should be new limits within which the great experimental simplification of miserly behavior can still be safely assumed as an approximation. Miserly utility, or utility with constant local risk aversion, seems to have been introduced by Pratt [43].

4. THE ELICITATION OF A RATE OF SUBSTITUTION

Some of the difficulties mentioned in the previous section tend to disappear when the investigation is confined to rates of substitution, of which probabilities are for this paper the prime example. First, the transactions envisaged are then moderate, almost by definition, so that typically the experiments that suggest themselves are not gradiose. Second, the satisfaction of one preference here has ideally no effect on the subject's other preferences, though in practice there can well be a conflict between keeping the transactions small enough to avoid important manifestations of saturation (or of nonlinear utility, as we say in connection with proability or expectation) and yet large enough to justify the subject's close attention. Possible remedies for this conflict, at least in probabilistic cases, were mentioned in Section 3.

Finally, by means of devices related to Marschak's method mentioned there, it is possible to present a subject with a single, relatively simple, economic choice in which it is to his interest to reveal any reasonable number of rates of substitution, as will now be explained for the case of a single rate.

Suppose the experimenter offers, once and for all, to buy some of a commodity at each possible price—more accurately price rate—so much at each rate. The subject will then have an incentive to satisfy the expressed demands of the experimenter at all rates higher than the subject's rate r but not those at lower rates, thereby revealing r.

In mathematical terms, the experimenter offers, for a certain non-negative "schedule of demands" f, to buy

$$\int_{x \leqslant \rho} f(\rho)d\rho \qquad (4.1)$$

units of the commodity at a total cost of

$$\int_{x \leqslant \rho} \rho f(\rho)d\rho \qquad (4.2)$$

for any number x named by the subject. From the subject's own viewpoint, his income from such a transaction, as a function of x, is

$$I(x;r) = -r \int_{x \leqslant \rho} f(\rho)d\rho + \int_{x \leqslant \rho} \rho f(\rho)d\rho$$
$$= \int_{x \leqslant \rho} (\rho - r)f(\rho)d\rho. \qquad (4.3)$$

Plainly, and in correspondence with the verbal argument that preceded this paragraph, the income $I(x;r)$ will be maximized when $x = r$, and for that x alone if $f(\rho)$ is positive near r on both sides. If, for example, $f(\rho) = \rho^{-3}$ for positive ρ, then $I(x;r) = (2x - r)/(2x^2)$, $I(x;r) - I(r;r) = -(x - r)^2/(2rx^2)$ $\leqslant 0$ for all positive r.

The mathematical analogy between this device for eliciting a rate and Marschak's device (discussed in paragraph 6 of Section 3) for eliciting a price is notable. The possibility that r is negative, in which case the subject regards the commodity as a nuisance, is not necessarily excluded but can often be eliminated from practical consideration. The experimenter might then be content, as in the example above, to define f only for positive prices. Similarly, the experimenter could with impunity attach 0 density to preposterously high prices.

With abbreviations, (4.3) reads

$$I(x;r) = b(x)r + c(x). \qquad (4.4)$$

The experimenter is therefore in effect offering to give the subject $c(x)$ units of cash and $b(x)$ units of the commodity for any number x chosen by the subject. The vital feature of the functions b and c, which is automatically assured if b and c are of the form implied by (4.3) with positive f, is that

$$I(x; r) = b(x)r + c(x) \leqslant b(r)r + c(r) = J(r), \qquad (4.5)$$

with equality if and only if $x = r$, where $J(r)$ has been introduced as an abbreviation for $I(r, r)$.

The sort of b and c induced by (4.4) with a positive f satisfies (4.5), but what can be said of the most general solution of (4.5)? In this problem, it is to be understood that x and r have a common convex range, such as all real numbers, the non-negative real numbers, or the real numbers between 0 and 1, inclusive or not. (A reader not familiar with convex sets and convex functions may find [47, Appendix 2] helpful here and elsewhere in the present article.) What (4.5) requires is that, for each fixed x, the function $I(x; r)$, which is linear in r, lies strictly below the function J except at x, where the linear function and J have the common value $I(x; x) = J(x)$. In short, $I(x; r)$ is, for each x, a linear function of support of J at the point x, and only there. This implies that J is a strictly convex function of r. Such a function J often has only one linear function of support at a given x, its tangent at x, but wherever J has a corner, it has more than one support.

Conversely, let J be any strictly convex function of r, well behaved at the endpoints of the range of r if there are any (that is, neither discontinuous nor vertical there), and let $I(x; r)$ be any support of J at x. This I evidently satisfies (4.5). (Examples of bad behavior at the endpoint 0 when the range is the non-negative reals: if $J(0) = 1$ and $J(x) = x^2$ elsewhere, then J is strictly convex but discontinuous and very much without a linear function of support at 0; if $J(x) = -x^{1/2}$, J is strictly convex and continuous but still has no linear function of support at 0.)

Any convex function J has, at each interior x, left and right derivatives $J_L(x)$ and $J_R(x)$. If $x < y$, $J_L(x) \leqslant J_R(x) \leqslant J_L(y) \leqslant J_R(y)$, and the middle inequality is strict if J is strictly convex. The slopes of the supports at x are the number between $J_L(x)$ and $J_R(x)$ inclusive. There is just one such slope if and only if this interval degenerates to a single number, which is then the derivative $J'(x)$. Since $b(x)$ is between $J_L(x)$ and $J_R(x)$, the function b is non-decreasing—strictly increasing if J is strictly convex. Consequently, except on and at most denumerable set of points, b is continuous, and except at just those points, b is the derivative of its indefinite integral. The points of continuity of b are the points of differentiability of J, and at those points, $J'(x) = b(x)$.

The function b determines J and c except for an additive constant k; thus

$$J(x) = \int\limits_{y \leqslant x} b(y)dy + k = b(x)x + c(x) + k. \tag{4.6}$$

If J is twice differentiable at x, then, as (4.6) makes clear, b and c are differentiable there, and therefore

$$b(c) = J'(x) = b(x) + b'(x)x + c'(x). \tag{4.7}$$

Whence

$$b'(x)x = -c'(x). \tag{4.8}$$

If, therefore, J' is absolutely continuous (that is, an indefinite integral of its own derivative where defined), and f denotes J''; then

$$I(x;r) = -r \int\limits_{x \leqslant \rho} f(\rho)d\rho + \int\limits_{x \leqslant \rho} \rho f(\rho)d\rho + dr + e, \tag{4.9}$$

where d and e are constants. Thus, under a rather mild regularity hypothesis, (4.3) is the general solution of (4.5), except for a relatively unimportant linear term. This term expresses the possibility of the experimenter's making the subject outright gifts (possibly negative) of d units of the commodity and e units of cash. Such a gift can have no rational influence on the subject's choice (if his utility is linear in the commodity and in cash), though it might play some practical role, such as insuring that an experimental subject will in net be paid, not penalized, for his cooperation.

Though the integral approach to the problem is of some interest and is not of seriously limited generality for one commodity (especially when extended by Stieljes-like integrals), it does not, over all, seem as useful and informative as (4.5) and will not be central to this article.

A convenient form for $I(x;r)$ is often

$$\begin{aligned} I(x;r) &= J(x) + b(x)(r - x) \\ &= J(x) + J'(x)(r - x), \end{aligned} \tag{4.10}$$

where the final line is applicable only if $J'(x)$ exists.

The loss to the subject if he (irrationally) replies with x when his rate is actually r is

$$\begin{aligned} L(x;r) &= I(r;r) - I(x;r) \\ &= J(r) - I(x;r) \\ &= J(r) - J(x) - b(x)(r - x) \\ &= J(r) - J(x) - J'(x)(r - x), \end{aligned} \tag{4.11}$$

where once more the final line is applicable only where the derivative $J'(x)$ exists.

The function L, and equivalently I, has an easily derived and useful monotonicity [6, p. 43; 30, p. 44], according to which it not only pays to choose x equal to r but to keep any unavoidable discrepancy small. Namely, if x is between r and z, then $L(x; r) \leqslant L(z; r)$, with strict inequality if J is strictly convex and $x \neq z$. Since L is non-negative and b is nondecreasing, the following identity makes this evident.

$$L(z; r) - L(x; r)$$
$$= J(x) - J(z) - b(z)(r - z) + J(x)(r - x)$$
$$= J(x) - J(z) - b(z)(x - z) + [b(z) - b(x)](x - r)$$
$$= L(z; x) + [b(z) - b(x)](x - r). \tag{4.12}$$

5. SOME SPECIAL FORMS FOR THE ELICITING FUNCTION

5.1 Loss as a Function of Discrepancy; $L(x; r) = H(x - r)$

One suggestive, and possibly desirable, way of limiting the choice of the convex function J is to require the subject's loss $L(x; r)$, if he replies with x when r is his true rate, to be a function only of his descrepancy $x - r$. Let us then investigate the consequences of supposing that $L(x; r) = H(x - r)$ for some function H, non-negative and 0 at 0 but not 0 everywhere. As the remaining paragraphs of this subsection are devoted to showing, this condition is so restrictive that $L(x; r)$ must be of the form $k(x - r)^2$ for some positive constant k. (This result has independently been derived with slightly less generality by Brown [6].)

According to (4.11),

$$H(x - r) = J(r) - J(x) - b(x)(r - x). \tag{5.1}$$

To see the sort of implication latent in (5.1), suppose for the moment that b and J are defined for all sufficiently small $|x|$ and $|r|$; as is rather intuitive and as will later be explicitly shown, this entails no real loss of generality. Under the simplifying assumption, an instance of (5.1) for $|x|$ and $|r|$ small is

$$H(x - r) = H((-r) - (-x))$$
$$= J(-x) - J(-r) - b(-r)(r - x). \tag{5.2}$$

Therefore

$$J(r) + J(-r) + b(-r)r - b(-r)x$$
$$= J(x) + J(-x) - b(x)x + b(x)r. \tag{5.3}$$

In particular,

$$J(r) + J(-r) + b(-r)r = 2J(0) + b(0)r, \tag{5.4}$$

$$2J(0) - b(0)x = J(x) + J(-x) - b(x)x. \tag{5.5}$$

By means of (5.4) and (5.5), simplify (5.3) thus,

$$\begin{aligned} 2J(0) + b(0)r - b(-r)x \\ = 2J(0) - b(0)x + b(x)r; \end{aligned} \tag{5.6}$$

or

$$[b(x) - b(0)]r = [b(0) - b(-r)]x, \tag{5.7}$$

so $b(x)$ is linear in x, at least for $|x|$ small.

To see that b is linear near every z in the interior of its domain of definition, and therefore linear throughout the interior, let $\hat{x} = x - z$, $\hat{r} = r - z$, $\hat{b}(u) = b(z + u)$, and $\hat{J}(u) = J(z + u)$. Since $(x - r) = (\hat{x} - \hat{r})$, the circumflexed functions and variables also satisfy (5.1), so $\hat{b}$ is linear for $|\hat{x}|$ small, that is, b is linear near z, as asserted, and consequently linear everywhere, with the (temporarily) possible exception of the endpoints of the domain of definition.

Armed with this information and the fact that H is non-negative but not identically 0, a reader familiar with the theory of the Cauchy-Hamel equation [1, Section 2.1] could easily show (from (5.1) alone) that $H(x - r) = k(x - r)^2$ for some positive k. But it is more elementary to recall that $b(x) = 2kx + l$ is the slope of a line of support of J at x. Since this slope is continuous in x, J is differentiable, and according to (4.6),

$$J(x) = kx^2 + lx + m, \tag{5.8}$$

with some positive k, and according to (4.10),

$$\begin{aligned} I(x;r) &= J(x) + J'(x)(r - x) \\ &= kx^2 + lx + m + (r - x)(2kx + l) \\ &= (2kx + l)r - kx^2 + m. \end{aligned} \tag{5.9}$$

And indeed, for any J of the form (5.8), according to (4.11),

$$\begin{aligned} L(x;r) &= J(r) - I(x;r) \\ &= kr^2 + lr + m - \left\{ [2kx + l]r - kx^2 + m \right\} \\ &= k(x - r)^2, \end{aligned} \tag{5.10}$$

as anticipated.

(The possibility that the linearity of b might fail at the endpoints of the interval of definition has been left open but can easily be removed by means of (5.1) and what has now been proved.)

5.2 Symmetry; $L(x; r) = L(r; x)$

Another suggestive condition is that the loss for replying with x when r applies should be the same as that for replying with r when x applies; that is, $L(x; r) = L(r; x)$. This condition too leads to (5.8), (5.9), and (5.10).

Briefly, the demonstration is as follows.

$$
\begin{aligned}
0 &= L(x; r) - L(r; x) \\
&= [b(r) + b(x)](x - r) + 2[J(r) - J(x)].
\end{aligned}
\tag{5.11}
$$

Consider the two equations that result from (5.11) on replacing (x, r) first by (r, z) and then by (z, x), and add all three equations to conclude that

$$
b(x)(z - r) + b(r)(x - z) + b(z)(r - x) = 0.
\tag{5.12}
$$

Therefore $b(x)$ is linear, and the passage to (5.8), (5.9), and (5.10) can be made as before. Or it can be made thus. When the form $b(x) = 2kx + 1/2$ is substituted into (5.11), $J(x)$ is seen to be of the form $kx^2 + lx + m$.

Quadratic J's are, then, not only the simplest convex functions from the algebraic point of view; they are characterized also by the condition that $L(x; r)$ is a function of $x - r$ and by the condition that $L(x; r) = L(r; x)$.

5.3 Ratio Discrepancy; $L(x; r) = H(r/x)$

When x ranges over the positive real numbers, the possibility that $L(x; r)$ is of the form $H(r/x)$ might be interesting. This condition too is very restrictive; it implies that J, I, and H are of the compatible forms:

$$
J(r) = m + lr - k \log r,
\tag{5.13}
$$

$$
I(x; r) = (m + k - k \log x) + (l - k/x)r,
\tag{5.14}
$$

$$
H(u) = k(u - 1 - \log u),
\tag{5.15}
$$

with k positive. The demonstration occupies the rest of this subsection.

To begin with,

$$
H(r/x) = J(r) - J(x) - J'(x)(r - x),
\tag{5.16}
$$

except for the at most denumerable set of x where $J'(x)$ is not well defined. Therefore, as is seen on differentiating with respect to r,

$$(1/x)H'(r/x) = J'(r) - J'(x), \tag{5.17}$$

whenever both derivatives on the right exist. Since these two derivatives do exist except on an at most denumerable set, a sequence of conclusions follows one after another: H is differentiable everywhere; so is J; (5.16) and (5.17) hold without exception; (5.16) can be differentiated on both sides with respect to x;

$$-(r/x^2)H'(r/x) = -(r - x)J''(x); \tag{5.18}$$

J is twice differentiable everywhere; and

$$rJ'(r) - rJ'(x) = x(r - x)J''(x). \tag{5.19}$$

Regarded as a differential equation for J in r, (5.19) easily implies (5.13); (4.10) then implies (5.14); and (4.11) implies (5.15). Since J as defined by (5.13) is strictly convex and results in (5.14) and (5.15), the system is indeed compatible.

5.4 Attempted Generalization; $L(x; r) = H(g(r) - g(x))$

Now consider the seemingly rather general possibility that $L(x; r)$ is of the form $H(g(r) - g(x))$. The special cases $g(x) = -x$ and $g(x) = \log x$ have already been considered. What other functions g are compatible with some strictly convex J? I am content here to examine the question under the simplifying assumption that g is differentiable and has a strictly positive derivative on the domain of J.

The derivation of (5.19) can be recapitulated to conclude that

$$\frac{J'(r)}{g'(r)} = \frac{(r - x)}{g'(x)} J''(x) + \frac{J'(x)}{g'(r)}. \tag{5.20}$$

Since the right side of (5.20) cannot change with x, and since $J'(x)$ is not constant, $1/g'(r)$ must be linear in r.

If this linear function is constant, it can without loss of generality be taken to be 1, with return to (5.8), (5.9), and (5.10). Incidentally, (5.20) specialized by setting $g' \equiv 1$ provides thus an alternative route to (5.8), (5.9), and (5.10).

If $1/g'(r)$ is linear but not constant, then it is to all intents and purposes of the form $(r - z)$ for some z below the domain of J or of the form $(z - r)$ for some z above the domain of J. In the first case, for example, $g(r) = \log (r - z) + \text{const.}$; and $L(x; r)$ is of the form $G((r - z)/(x - z))$, which is virtually the form that led to (5.13), (5.14), and (5.15). The introduction of g has therefore led to no really new forms of I, J, and L.

For J defined on the unit interval, it might seem interesting to seek $L(x; r)$ of the form $H(r(1 - x)/(1 - r)x)$. But that would imply the existence of a $g(r)$ of the form $\log r - \log (1 - r)$, which has been shown to be impossible.

6. PROBABILISTIC INTERPRETATION

6.1 Application to a Probability as a Rate of Substitution

Suppose that an experimenter, in an effort to elicit the probability p that a subject associates with an event D, invites the subject to choose a number x and promises to pay him $Y(x)$ in case D obtains and $Z(x)$ in case D does not obtain. For what pairs of functions Y and Z will it be to the subject's interest to choose x equal to p?

If p is interpreted as the subject's rate of substitution of dollars for the commodity consisting of dollars that are contingent on D, then the experimenter is in effect offering the subject $Z(x)$ dollars in cash and $Y(x) - Z(x)$ units of the commodity. The worth of such a gift to the subject is

$$\begin{aligned} I(x; p) &= [Y(x) - Z(x)]p + Z(x) \\ &= Y(x)p + Z(x)(1 - p), \end{aligned} \tag{6.1}$$

which is an instance of (4.4) with p, $Y(x) - Z(x)$, and $Z(x)$ playing the roles of r, $b(x)$, and $c(x)$.

Therefore Y and Z accomplish the objective if and only if

$$\begin{aligned} J(p) &= Y(p)p + Z(p)(1 - p) \\ &= Y(p) - (Y(p) - Z(p))(1 - p) \\ &= Z(p) + (Y(p) - Z(p))p \end{aligned} \tag{6.2}$$

is a strictly convex function of p, and $I(x; p)$ is in p a linear function of support of J at x. At values of p where J is differentiable, according to (6.1),

$$J'(p) = Y(p) - Z(p), \tag{6.3}$$

so at such values, according to (6.2) and (6.3),

$$Y(p) = J(p) + (1 - p)J'(p) \tag{6.4}$$

$$Z(p) = J(p) - pJ'(p). \tag{6.5}$$

The loss entailed by choice of x when p applies is

$$L(x;p) = [Y(p) - Y(x)]p + [Z(p) - Z(x)](1 - p), \qquad (6.6)$$

which is ordinarily—that is, if J is differentiable—

$$L(x;p) = J(p) - J(x) - J'(x)(p - x), \qquad (6.7)$$

as in (4.11).

The conditions that J is quadratic, that $L(x;p)$ is a function of $p - x$, and that $L(x; p) \equiv L(p; x)$ are, according to Section 5, all equivalent. In this special case, (6.4) and (6.5) can be put in the suggestive forms

$$Y(p) = m' - k(1 - p)^2, \qquad (6.8)$$

$$Z(p) = m - kp^2, \qquad (6.9)$$

which correspond to

$$J(p) = m'p + m(1 - p) - kp(1 - p). \qquad (6.10)$$

In any real application, p is between 0 and 1, and it might seem natural to confine the range of choice of x to the interval from 0 to 1 inclusive. However, if x is not so confined, a subject who chooses $x < 0$ or $x > 1$ exposes himself to utterly unnecessary loss for any strictly convex J. In fact, according to the monotonicity pointed out in the final paragraph of Section 4, for any p in the interval $[0, 1]$, $L(x;p) > L(1;p)$ if $x > 1$, and $L(x;p) > L(0;p)$ if $x < 0$.

6.2 Application to an Expectation as a Rate of Substitution

Since the probability of an event D is the expectation of the same D regarded as an indicator, extension of the method of eliciting probabilities just discussed to the elicitation of the personal expectation r of any random quantity V is to be anticipated. And this is indeed straightforward. If the experimenter offers to pay the subject $b(x)V + c(x)$ for any x chosen by the subject, the worth to the subject of choosing x is

$$I(x;r) = b(x)r + c(x), \qquad (6.11)$$

so the loss for choosing x instead of r is

$$L(x;r) = [b(r) - b(x)]r + [c(r) - c(x)]. \qquad (6.12)$$

For this to be positive if and only if x differs from r means, as shown in Section 4, that $J(r) = b(r)r + c(r)$ is strictly convex and $b(x)r + c(x)$ is in r a linear function of support of J at x.

If, for example, the subject feels certain that $V \leqslant \alpha$ for some constant α, his $r = E(V)$ cannot exceed α. And in fact it is to his advantage to choose α rather than any x larger than α, as can be verified thus. The difference in worth between these two choices is

$$[b(\alpha) - b(x)] V + [c(\alpha) - c(x)] = L(x; V) - L(\alpha; V). \qquad (6.13)$$

In view of the monotonicity mentioned at the end of Section 4, (6.13) is therefore negative if $x > \alpha$. Thus whatever the actual value of V may be—not greater than α—the subject will receive more for choosing α than for choosing any larger number.

6.3 The Most General Eliciter of an Expectation

Is $b(x)v + c(x)$ the only possible form for a function $S(x; v)$ for which

$$E(S(x; V)) < E(S(r; V)) \qquad (6.14)$$

for all x different from $r = E(V)$? The answer cannot quite be "yes," because to any such S it is clearly possible to add $f(V)$ with f any function for which $E(f(V))$ is finite for the class of distributions envisaged for V. But little if any further extension is possible if (6.14) is to hold for a reasonably large class of distributions for V, as the next four paragraphs demonstrate. Even this slight extension is nugatory in case V is the indicator of an event, that is, in case $E(V)$ is a probability, because when V takes only two values, any function f is linear on that pair of values.

To begin with, let $S(x; v)$ be defined for all x and v in an interval, and let (6.14) hold for every V that is subject to a 2-point distribution in that interval. Specialized to a V that takes the values v and v' with probabilities p and $\bar{p} = 1 - p$, (6.14) takes the form:

$$pS(x, y) + \bar{p}S(x; v') < pS(pv + \bar{p}v'; v) + \bar{p}S(pv + \bar{p}v'; v'), \qquad (6.15)$$

unless $x = pv + \bar{p}v'$.

For x interior to the interval of definition and S differentiable in x,

$$p \frac{\partial}{\partial x} S(x; v) + \bar{p} \frac{\partial}{\partial x} S(x; v') = 0 \qquad (6.16)$$

if $x = pv + \bar{p}v'$. That is, if $v \leqslant x \leqslant v'$ and the derivatives exist at x, then

$$(v' - x) \frac{\partial}{\partial x} S(x; v) = (v - x) \frac{\partial}{\partial x} S(x; v'). \tag{6.17}$$

Therefore, if S is everywhere differentiable with respect to x (more generally, if S is absolutely continuous in x for each v), then $S(x; v)$ is of the form $b(x)v + c(x) + f(v)$ as anticipated. This conclusion obtains even without any differentiability assumption, as will be shown in the next paragraph for those who share an interest in such points.

According to (6.15), the function

$$(v' - x)S(x; v) + (x - v)S(x; v') \tag{6.18}$$

is convex in x and has

$$S(x; v') - S(x; v) \tag{6.19}$$

among its slopes of support at x. Therefore the functions (6.18) and (6.19) are of bounded variation in x, whence so is $S(x; v)$ for each v. Let S_0 and S_1 be the singular and the absolutely continuous parts of S with respect to x, rendered unique by the convention that $S_0(x_0; v) = 0$ for some x_0 and for arbitrary v. Since (6.18), being convex, is absolutely continuous, $S_0(x; v) = (v - x)S_0(x; v')/(v' - x)$ and is therefore of the form $b_0(x)v + c_0(x)$. (Now, the argument employed when S was assumed to be absolutely continuous applies almost unmodified to S_1, the absolutely continuous part of S.)

An important and widely known example is

$$S(x; v) = -(x - v)^2 = -x^2 + 2xv - v^2.$$

For it is central to the theory of least squares that this S satisfies (6.14) according to the following familiar calculation.

$$\begin{aligned} E[(x - V)^2] &= E\left\{[(x - E(V)) - (E(V) - V)]^2\right\} \\ &= (x - E(V))^2 + E[(E(V) - V)^2]. \end{aligned} \tag{6.20}$$

7. DECISION THEORY

A different approach to the problem of determining those functions Y and Z that encourage a subject to reveal his true p is implicit in statistical decision theory, as will now be explained.

Imagine a person in an economic situation in which he is free to choose one of several acts and knows that if he chooses the act a he will receive a payment of $Y(a)$ dollars in case the event D obtains and $Z(a)$ dollars in case

D does not obtain. If the person's probability for D is p, the worth of the act a to him is

$$I(a; p) = Y(a)p + Z(a)(1 - p),\qquad (7.1)$$

a linear function of p. The worth to him of the situation in which he is free to choose among a finite set of acts a is therefore the function

$$J(p) = \max_a I(a; p).\qquad (7.2)$$

Indeed, any act a for which the maximum is attained is worth $J(p)$ to him and no available act is worth more. The function $J(p)$ is evidently convex.

When an infinite set of acts is envisaged, little changes, especially if the set is such that the maximum required by (7.2) is attained. With an unrestricted infinite set of acts, J can be an arbitrary continuous, convex function. (If the end points $p = 0$ and $p = 1$ are to be included and if the maximum is to be attained, J cannot be permitted to be vertical at those points.) If J is strictly convex, a person choosing a support of J will reveal his value of p; for no linear function supports a strictly convex function at more than one point.

The original problem of characterizing functions Y and Z that will elicit the person's true p can now be envisaged thus. For each number x, say in the the interval from 0 to 1, an act is made available to the subject that will pay him $Y(x)$ dollars in case D obtains and $Z(x)$ dollars otherwise. For what functions Y and Z will it be optimal for the person to choose x equal to his p, no matter what p may be? As the preceding paragraph shows, it is necessary and sufficient that there be some strictly convex J for which $Y(x)p + Z(x) (1 - p)$ is a line of support at x; and this was the main conclusion of Section 6.

8. AN ALTERNATIVE DEFINITION OF PROBABILITY

The preceding section suggests an alternative way to define personal probability and to argue that personal probability exists. This avenue has been pointed out and explored by de Finetti, for example, in [13, 17].

Imagine in fact that the very notion of the probability p attached by a person to the event D has not yet been defined, and consider a person confronted with the choice among all those acts a for which the linear functions $I(a; p)$ nowhere exceed a specified convex function $J(p)$ (without vertical endpoints). Suppose that there is one such act a to which he prefers no other. It would be unreasonable for this a not to be one for which $I(a; p)$ is a line of support of $J(p)$ at some point p_0. For otherwise there would be an a' for which $I(a'; p)$ is a line of support of J and is parallel to $I(a; p)$, in which case $Y(a') > Y(a)$ and $Z(a') > Z(a)$, so that a' is clearly superior to a.

If J is strictly convex, the unique p_0 for which $I(a; p)$ is a line of support of J can be defined as the person's personal probability of D. But this raises three important questions: Could a different fucntion J^* lead to a different p_0? Does this definition lead back to the idea of probability as a rate of substitution? Is probability as thus defined a probability measure?

To progress toward answering the first two questions, suppose the person participates simultaneously in two decision problems of the kind under discussion, one determined by J_1 and one determined by J_2, where J_1 and J_2 are convex but not necessarily strictly convex. Assume that he acts in each of the two component problems as he would if he were faced with that problem alone.

(This assumption is not altogether unobjectionable; for it may imply that the person's utility function is linear in money. But such linearity assumptions are made almost throughout the present paper and are presumably tolerable if only moderate sums of money are involved. In the purely mathematical formulation of the decision problems no precautions have been, or need be, taken to keep these sums moderate, but it is fairly clear how such precautions could be taken in applying the theory or how the devices mentioned at the end of Section 3 might be invoked. I do not attempt great caution about this point, because my object in this section is only to touch briefly on an approach to personal probability that may be more suggestive and more practical in some respect than such formally more rigorous approaches as those reported on in [47, Ch. 3].)

The linear function of support $I_1(a_1; p)$ chosen for J_1 and the linear function of support $I_2(a_2; p)$ chosen for J_2 should support J_1 and J_2 respectively at some common point p^*, as will be argued. The person, in choosing a_1 and a_2, has in effect chosen an act a for which $I(a; p) = I_1(a_1; p) + I_2(a_2; p)$. Some linear function of support $I^*(p)$ of $J = J_1 + J_2$ at some point p^* is either the same as, or everywhere higher than, $I(a; p)$. Since, as is not hard to see, I^* can be represented as the sum of linear functions of support of J_1 and J_2 at p^*, I^* represents an act (that is a choice of a_1 and a_2) that was available to the person. Therefore his choice of a_1 and a_2 is discredited, or seen to be inadmissible, unless $I^*(p) = I(a; p)$ for all p, which is impossible unless $I_1(a_1; p^*) = J_1(p^*)$ and $I_2(a_2; p^*) = J_2(p^*)$ as asserted. This conclusion will yield affirmative answers to the first two questions.

First, as is now evident, two strictly convex functions cannot lead to different values of p if the linear-utility assumption holds and if the person is coherent.

Second, suppose p^* has been determined by means of a strictly convex function J_1, and consider also the decision problem defined by the broken-line convex function J_r with $J_r(p) = \max\ (rq,\ pq)$, where r is a positive fraction and q is a positive number. According to what was proved in the paragraph before last, the person will prefer q dollars contingent on D to rq dollars outright if $r > p^*$ and vice versa if $r < p^*$, so p^* is indeed a rate of substitution.

A certain approach to the third question, whether probability as here defined is a probability measure, is best postponed to the next section. But it can be argued now that since personal probabilities defined as rates of substitution constitute a probability measure, the same must be true for the equivalent new definition.

De Finetti [13] has shown how the approach of this section applies to conditional probability.

9. THE SIMULTANEOUS ELICITATION OF SEVERAL RATES

9.1 Vector Rates

When several rates r_s are concerned, each can be independently elicited by means of a function $I_s(x_s;\ r_s)$ as in Section 4, but there are other possibilities.

Generalizing from Section 4, let $r = \{r_s\}$ be a finite sequence of rates, which can advantageously be thought of as a vector. (Infinite sequences or even functions on an arbitrary domain might also have a useful interpretation, for instance, as distribution functions or densities. Vectors without any interpretation as functions could also be handled and might have applications. Extensions of this sort are presented by Hendrickson [30].) Let $x = \{x_s\}$ be a sequence of possible numerical responses of a subject, a response vector. The experimenter's aim is to provide the subject with an incentive to choose $x = r$ by offering to give $c(x)$ units of cash and $b_s(x)$ units of the sth commodity for whatever x is chosen by the subject.

The experimenter may already know something about r. For example, in the important case of probabilities, he knows that each probability is between 0 and 1; and when the r_s are probabilities of the elements of a partition, he knows that they add up to 1. In other cases, it will often be known that each $r_s > 0$. For still one more example, the rate for a certain high quality commodity may be plainly higher than that of certain other commodities.

It may be to the experimenter's advantage to exclude all, or at least some, unreasonable values of x. In particular, he might confine acceptable responses x to some convex set K—a scheme that gives adequate flexibility for the examples just mentioned. For the moment, assume that K has at least one extrinsic interior point z, that is, a point for which $z + \Delta r = \{z_s + \Delta r_s\}$ is in K for all sufficiently small vectors Δr. This temporary simplification excludes at least one important example, that of probabilities p_s so constrained that $\Sigma p_s = 1$.

With the exception of the expression of I by integrals, the whole of Section 4 will now easily be paraphrased with sets, or bundles, or commodities (or, more abstractly, vector commodities) playing the role that a single commodity played there.

Though the experimenter does not initially know the vector r, he can in effect offer the subject the income $I(x; r)$ for choosing the vector x, where

$$
\begin{aligned}
I(x; r) &= \Sigma_s b_s(x) r_s + c(x) \leqslant \Sigma b_s(r) r_s + c(r) \\
&= I(r; r) = J(r),
\end{aligned}
\tag{9.1}
$$

with equality if and only if $x = r$.

According to (9.1), $I(x; r)$ is, for each x, a strict linear function of support at x for the function J. Therefore, J is strictly convex, and virtually all strictly convex functions do lead to at least one I. If K has no boundary points—and it is seldom if ever important to include boundary points in K—then every strictly convex function will serve. If K does have boundary points, then discontinuities at the boundary points and milder sorts of misbehavior incompatible with linear functions of support at the boundary points are to be excluded.

If J is differentiable at the nonboundary point x of K or, equivalently, if the $b_s(x)$ are continuous at x, then

$$
b_s(x) = \frac{\partial}{\partial x_s} J(x),
\tag{9.2}
$$

and

$$
c(x) = J(x) - \Sigma x_s \frac{\partial}{\partial x_s} J(x).
\tag{9.3}
$$

If the $b_s(x)$ are continuous for all such x, they obviously determine J through (9.2), except for an additive constant. (But continuity is not needed for the conclusion that J is determined except for an additive constant by the b_s.)

The loss incurred by the subject on choosing x when r applies is

$$
\begin{aligned}
L(x; r) &= [J(r) - J(x)] - \Sigma b_s(x)(r_s - x_s) \\
&= [J(r) - J(x)] - \Sigma (r_s - x_s) \frac{\partial}{\partial x_s} J(x),
\end{aligned}
\tag{9.4}
$$

the last line applying only where J is differentiable.

The condition that $J(x)$ be quadratic, that is, of the form

$$J(x) = \sum_{s,t} k_{s,t} x_s x_t + \Sigma_s 1_s x_s + m \tag{9.5}$$

is equivalent to the condition that $L(x; r)$ depends only on $x - r = \left\{ x_s - r_s \right\}$ and to the condition that $L(x; r) = L(r; x)$.

Much the same can be said even without assuming that K has extrinsic interior points, with slight differences because it may no longer be possible to change one x_s without changing other components of x. Equation (9.1) and the paragraph following it remain in force, but it now may be possible to change the individual $b_s(x)$ without really affecting I. The meaning of (9.1) and (9.4) is therefore better conveyed by

$$I(x; r) = b(x)(r) + c(x) \leqslant b(r)(r) + c(r) = J(r), \tag{9.6}$$

$$L(x; r) = J(r) - J(x) - b(x)(r - x), \tag{9.7}$$

where for each x, $b(x)(r)$ is linear and homogeneous in the vector r and $I(x; r)$ is, as always, a linear function of support of J at x.

9.2 Vector Expectations

The case of a sequence $V = \left\{ V_s \right\}$ of random numbers considered as commodities is important. A little more generally, V can be a random vector whose values lie in a finite dimensional convex set K. The random payment to the subject who chooses the vector x is

$$b(x)(V) + c(x), \tag{9.8}$$

which in V is linear and supports a strictly convex J at x. Where J is differentiable, (9.8) can be suggestively written as

$$J(x) + J'(x)(V - x). \tag{9.9}$$

As in Subsection 6.3, it can be asked whether there are functions $S(x; v)$ that elicit the expected value of a vector V other than S of the form $I(x; v) + f(v)$, where I is as in (9.6). The negative answer given in Subsection 6.3 is not hard to generalize, at least under generous regularity hypotheses.

The interesting and familiar instance of quadratic S mentioned in Subsection 6.3 has a hardly less familiar extension to the present, multidimensional, situation. For if S is a homogeneous, strictly negative-definite, quadratic function of $x - v$, then

$$E[S(x - V)] = S(x - E(V)) + E[S(V - E(V))] > E[S(V - E(V))] \quad (9.10)$$

if $x \neq E(V)$.

A subject who feels sure that V is in a specified closed convex subset K^* of K will have $E(V)$ in K^* and therefore ought not choose an x not in K^*. But, under suitable regularity hypotheses, such an x is also discredited in the deeper sense that there is a v in K^* for which $I(v; r) > I(x; r)$ for every r in K^*. This is demonstrated in the next paragraph, which is followed by one showing that some regularity hypothesis is indispensable.

Suppose that J is differentiable in K^* and that the infimum of $L(x; r)$ as a function of r in K^* is attained, say at $r = v$. Then, for all r in K^*,

$$\begin{aligned} L(x; r) &= J(r) - J(x) - b(x)(r - x) \\ &\geq J(v) - J(x) - b(x)(v - x) = L(x; v) \end{aligned} \quad (9.11)$$

$$\begin{aligned} 0 &\leq L(x; r) - L(x; v) \\ &= [J'(v) - b(x)](r - v) + o(r - v). \end{aligned} \quad (9.12)$$

Therefore, $[J'(v) - b(x)](r - v) \geq 0$ for all r in K^* sufficiently close to v; but if such a linear inequality holds for r in K^* close to v, it holds for all r in K^*. Put the facts together, thus.

$$\begin{aligned} L(x; r) - L(v, r) &= J(v) - J(x) - b(x)(r - x) + J'(v)(r - v) \\ &= [J'(v) - b(x)](r - v) + L(x; v) \\ &\geq 0, \end{aligned} \quad (9.13)$$

as asserted.

(In one dimension, as has been seen in Subsection 6.2, no regularity hypothesis is required for this conclusion, but the following counterexample shows that some such hypothesis is needed in two and more dimensions. Let $J(x, y) = |x| + |y| + \epsilon(x^2 + y^2)$ for some small positive ϵ. This J is strictly convex over the whole plane, and I can be consistently defined thus.

$$I(x, y; r, s) = J(x, y) + (\text{sgn } x + 2\epsilon x)(r - x) + (\text{sgn } y + 2\epsilon y)(s - y).$$

Let K be the convex set

$$\left\{ x, y : 1 \leq 2x + y; x \leq 2, y \leq 3 \right\}.$$

Then $I(0, 0; r, s) = 0$ for all (r, s), but, for each (x, y) in K,

$$\inf_{(r, \delta) \in K} I(x, y; r, s) \leq -1 + 0(\epsilon).)$$

The significance of the general conclusion in case of regularity is particularly vivid if J is quadratic. For then, in terms of the Euclidean distance associated with the quadratic function J, (9.13) says that the point v in K^* closest to x is closer to each point of K^* than x is. This is geometrically rather evident and is easy to prove directly or as an instance of the general argument.

To apply the argument about K^*, if the person is sure that a linear equality or inequality is satisfied by several random variables $V_1, \cdots, V_n$, he will expose himself to needless loss if the $x_1, \cdots, x_n$ with which he responds to any regular J do not satisfy this equality or inequality. In particular, if some V_s is sure to be 1, the person should choose $x_s = 1$; if V_s is the indicator of an event, he should choose x_s between 0 and 1; if V_s is the indicator of the union of two disjoint events of which V_j and V_k are the indicators, he should so choose that $x_s = x_j + x_k$. As these remarks show, the program for defining probabilities in Section 8 does lead to a probability measure for any person who does not blunder. (More completely, the remarks do lead directly to the anticipated conclusion when the program is applied to any J that is regular. But, as was shown in Section 8, all strictly convex J elicit the same probabilities.)

9.3 Probability Distributions as Vector Expectations

Consider now the case in which the V_s $(s = 1, \cdots, n)$ are the indicators of the elements of a partition $\{D_s\}$. All previous literature on rate elicitation seems to be confined to this important case. (A few references on the method not mentioned elsewhere in this article are [7, 42, 54].)

Since it is (especially patently for regular J) wasteful in the present case for the person to choose any $x = \{x_s\}$ other than a probability distribution (that is, an x for which $x_s \geqslant 0$ and $\Sigma x_t = 1$), K can for many applications be taken to be the simplex of all probability distributions p. If $J(p)$ is a differentiable function on K, the person who chooses x will receive $I(x; V)$, which can, in a somewhat figurative but not unnatural notation, be written,

$$
\begin{aligned}
I(x; V) &= J(x) + J'(x)(V - x) \\
&= J(x)\ \Sigma_t J_t'(x)(V_t - x_t);
\end{aligned} \tag{9.14}
$$

this is figurative because x cannot be varied one coordinate at a time and remain in the simplex. In case D_s is the event in the partition that actually obtains, this is,

$$
I(x; V) = J(x) - \Sigma_t J_t'(x)x_t + J_s'(x). \tag{9.15}
$$

The numbers $J_s'(x)$ here are not necessarily derivatives and are determined only up to an additive constant. (They can of course be so chosen that $\Sigma J_t'(x) = 0$, but the symmetry of that particular choice ought not to be invested with much importance.) If J is originally defined not only on the simplex K but, say, for all positive n-tuples of numbers, then the $J_s'(x)$ can be taken to be $\partial J(x)/\partial x_s$, which would not be meaningful on K alone.

One suggestive choice for J, whether on K or on a larger set, is $J(x) = \Sigma x_t^2$. In this case, (9.15) on K becomes

$$
\begin{aligned}
I(x; V) &= -\Sigma_t x_t^2 + 2x_s \\
&= -(1 - x_s)^2 - \Sigma_{t \neq s} x_t^2 + 1 \\
&= -\Sigma_t (V_t - x_t)^2 + 1,
\end{aligned}
\tag{9.16}
$$

that is, 1 minus the square of the usual Euclidean distance from x to the vector $(0, 0, \cdots, 1, \cdots, 0, 0)$ representing the D_s that actually obtains. The final constant 1 is of course not very important in (9.16). Specialized to a single event D (with indicator V) and its complement, (9.16) without its 1 becomes

$$
\begin{aligned}
I(x; V) &= -[V - x]^2 - [(1 - V) - (1 - x)]^2 \\
&= -2(V - x)^2 \\
&= \begin{cases} -2(1 - x)^2 & \text{if } D \text{ obtains} \\ -2x^2 & \text{if } D \text{ does not obtain.} \end{cases}
\end{aligned}
\tag{9.17}
$$

If each of the events D_s is treated separately according to (9.17), the net effect is practically the same as that of using $J = \Sigma x_t^2$; it is exactly the same as using $2J - 2$.

The most general quadratic function of n variables is

$$
J(x) = \Sigma_{s,t} k_{st} x_s x_t + \Sigma_s l_s x_s + m.
\tag{9.18}
$$

To this J corresponds

$$
I(x; V) = J(V) - \Sigma_{s,t} k_{st}(V_s - x_s)(V_t - x_t).
\tag{9.19}
$$

If further, J is symmetric in the x_s, (9.18) specializes to

$$
J(x) = k' \Sigma_s x_s^2 + k'' (\Sigma_s x_s)^2 + l \Sigma_s x_s + m,
\tag{9.20}
$$

which is convex if and only if $k' > 0$ and $k' + nk'' > 0$. For this J,

$$I(x; V) = J(V) - k' \Sigma_s (V_s - x_s)^2$$
$$- k'' [\Sigma_s (V_s - x_s)]^2. \tag{9.21}$$

If the V_s are the indicators of a partition, $\Sigma V_s^2 = \Sigma V_s = 1$; so $J(V) = k' + k'' + l + m$, which simplifies (9.21). If further $\Sigma x_s = 1$ as it "ought to," then (9.21) simplifies to

$$(k' + k'' + l + m) - k' \Sigma_s (V_s - x_s)^2, \tag{9.22}$$

which is not essentially different from (9.16).

9.4 Separated Income for Distributions

It is somewhat attractive to seek functions J on the simplex K such that when D_s obtains the income of the subject will depend on x_s alone. The pertinent facts have been known for some time, a review of them here without any supplementary hypotheses may be useful. (An early reference is [37], which attributes the main fact to Andrew Gleason.)

If $n = 2$, the condition is plainly vacuous, but for $n > 2$ it is very severe—strictly speaking, not quite attainable. What is wanted is n functions f_s on the interval from 0 to 1 (inclusive if possible) such that, for any distribution p,

$$\Sigma_t f_t (x_t) p_t < \Sigma_t f_t (p_t) p_t \tag{9.23}$$

for every distribution x different from p. As will be shown, after preliminary discussion, if this is to hold for all p with each p_s strictly positive, then

$$f_s(z) = k \log z + l_s, \tag{9.24}$$

for some $k > 0$. Whence

$$J(p) = k \Sigma_s p_s \log p_s + \Sigma l_s p_s; \tag{9.25}$$

$$L(x; p) = k \Sigma_s p_s \log \frac{p_s}{x_s}. \tag{9.26}$$

Of course (9.23) and (9.24) are not compatible with any assignment of a finite value for $f_s(0)$.

The loss (9.26) is a constant multiple of what is sometimes called the information of the distribution p with respect to the distribution x. The scheme of elicitation implied by (9.25) seems to have been suggested first by Good [26, p. 112], who confined himself to two-fold partitions, for which the uniqueness theorem is not relevant. An interesting application and discussion are given by Mosteller and Wallace [38, Section 4.9].

It is not hard to prove (9.24) once it is known that each f_s is sufficiently regular. The next three paragraphs are devoted to proving that regularity.

Apply (9.23), for $n > 2$, to distributions x and p of the special form $x = \left\{ yw, \bar{y}w, \bar{w}/(n-2), \cdots, \bar{w}/(n-2) \right\}$ and $p = \left\{ qw, \bar{q}w, \bar{w}/(n-2), \cdots, \bar{w}/(n-2) \right\}$, where $\bar{y} = 1 - y$, $\bar{q} = 1 - q$, $\bar{w} = 1 - w$, and $0 < y, q, w < 1$.

$$f_1(yw)q + f_2(\bar{y}w)\bar{q} < f_1(qw)q + f_2(\bar{q}w)\bar{q} \qquad (9.27)$$

if $y \neq q$. The left side of (9.27) is therefore, in q, a strict linear function of support at y of g_w, where

$$g_w(y) = f_1(yw)y + f_2(\bar{y}w)\bar{y}. \qquad (9.28)$$

Therefore, g_w is strictly convex in $(0, 1)$. So g_w is continuous in y and, except possibly on a denumerable set, differentiable in y.

Since $f_1(yw) - f_2(\bar{y}w)$ is a slope of support of g_w at y, it is locally of bounded variation, and the same can therefore be said of $f_1(yw) = g_w(y) + \bar{y}[f_1(yw) - f_2(\bar{y}w)]$ as a function of y. Therefore each f_s has at most a denumerable number of discontinuities and is differentiable almost everywhere.

Returning to (9.28) and recalling that g_w is continuous in y, it can be seen that if f_1 is discontinuous at z, then f_2 is discontinuous at $w - z$ for all w between z and 1; so f_1, and therefore all f_s, must be continuous everywhere. Therefore, g_w is differentiable at all y. Arguing as before, if f_1 is not differentiable at z, then f_2 is not differentiable at $w - z$. At last, we know that each f_s is differentiable at all z in $(0, 1)$. Let f_s' be the derivative of f_s.

The rest is straightforward:

$$\frac{d}{dy} g_w(y) = f_1(yw) - f_2(\bar{y}w)$$

$$= \frac{d}{dy} \left\{ f_1(yw)y + f_2(\bar{y}w)\bar{y} \right\} \qquad (9.29)$$

$$= w \left\{ f_1'(yw)y - f_2'(\bar{y}w)\bar{y} \right\}$$
$$+ f_1(yw) - f_2(\bar{y}w)$$

$$f_1'(yw)yw = f_2'(\bar{y}w)\bar{y}w. \qquad (9.30)$$

This means that $f_s'(z)z = k$ independently of s or z, which establishes (9.24).

10. PRACTICAL SIDELIGHTS

10.1 Scope of this Section

This article, which is largely about the mathematical aspects of the procedures for eliciting personal probabilities that are now often called proper scoring rules (and also admissible probability measurement procedures) would be misleading and incomplete without some discussion of their actual and potential applications. My preparation for that is inadequate; for I have done no practical work in the area nor even followed the practically oriented literature with energy and care. But it seems incumbent on me to mention, to the best of my knowledge, the sorts of applications that have been envisaged, some difficulties that threaten them, and criteria that might help in selecting among the plethora of proper scoring rules. The subject is a ramified one, and even a cursory survey demands considerable space.

One serious omission is discussion of the experiments that have been done on proper scoring rules, for which some key references are [12; 52, Sec. 3.4 and Ch. 10-11; 55; 56].

The applications thus far envisaged have tended to emphasize the elicitation of probabilities over that of other expectations. For this reason and for vividness, I focus here on probabilities, but extension of the ideas to other expectations will often be obvious and ought to be kept in mind.

10.2 The Uses of Opinions

Strictly proper scoring rules enable us, in principle, to discover people's opinions, so possible fields of application are brought to mind by asking why and when we are interested in opinions.

Often, we want to make use of the opinion of a person whom we regard as an expert. Does the weatherman think that it will rain, the doctor that we shall soon get well, the lawyer that it would be better to settle out of court, or the geologist that there might be lots of oil at the bottom of a deep hole? Most of the following subsections are concerned explicitly with the utilization of experts.

And often, we want to know a person's opinions in order to judge how well informed he is. Every academic examination can be viewed in that light. It is, therefore, interesting to explore the possible usefulness of proper scoring rules in academic examinations, and the final subsection returns to academic testing. This domain of application shades into that of trying to determine which among possible experts are most valuable for a given task by means of the relation between their past opinions and reality.

Since public-opinion polls are ostensibly concerned with finding out the opinions of the public, we might at first expect proper scoring rules to have important applications there, in harmony with the remarks of Haavelmo quoted in Section 1. This has, however, apparently never been suggested, partly perhaps because public-opinion polls are seldom so much concerned with the opinions of the public as with the preferences of the public. But investigators do occasionally want to know how firmly the public believes some as yet unresolved matter of fact, and in a few of these cases, proper scoring rules might have some role. Literally paying the participants in a poll on the basis of the accuracy of their predictions would of course ordinarily be infeasible, especially in the common case of casual, one-time participants. Yet scoring rules might be used in training panelists to assess their own probabilities, perhaps partly in paid practice sessions about immediately verifiable predictions. This would seem less farfetched should scoring rules come to be widely applied in the schools.

One domain of potential application of proper scoring rules is suggested not so much by the problem of obtaining the opinions of others as by the difficulty of obtaining our own. For it is by no means easy to elicit your own probabilities. Vagueness is a major obstacle, and your first reactions are often greatly modified when you reflect upon their implications. Those who have experimented on themselves and on others generally feel that frequent practice with proper scoring rules and with other probability elicitors helps a person to combat vagueness and to arrive more promptly and accurately at his personal probability. This proposition might be difficult to investigate experimentally and even seems difficult to state with precision, but its promise of benefit is great. (See, for example, [12, 59].)

10.3 Yes, No, or Maybe Is Not Enough

Traditionally, experts, except for turf experts, have not communicated their opinions in probabilities. The doctor says, "That child will soon be well and then he had better have his tonsils out." The geologist says, "That looks like a good place to drill." And, until recently, most of us were content when the weatherman said simply, "Rain tonight."

The importance of a system of communication in which experts express themselves in terms of genuine personal probabilities and in which those who utilize the opinions of experts—that is, all of us—are trained in understanding and using such probabilities was energetically underlined by Grayson [27]. In recent years, meteorologists have been announcing forecasts in terms of probability on some radio and television services, but not all who offer these probability forecasts think in terms of personal probability, and some seem to be very vague indeed about what they mean by probability in a forecast. The earliest known reference to proper scoring rules is by the meteorological statistician, Brier [5], and much of the current literature on proper scoring rules is inspired by meteorology, as in [21, 39, 40, 41, 46, 53, 58, 59, 60] and works cited in them.

Applying a proper scoring rule to obtain better opinions from an expert might mean, at one extreme, merely using the rule to keep score as a training device to give the expert a mild incentive to understand what probability means and to ask himself whether it is really his personal probabilities that he is reporting. At the opposite extreme, the scoring rule might be implemented by substantial cash payments. For example, an oil geologist willing to buy a $100 interest in a million dollar drilling investment is saying very clearly that to him, the expected revenue of the well is greater than the expected cost. More accurately, that is what he is saying if his expression of opinion is not affected by nonlinearities in his utility, possibly reflecting the nonlinearity of income after taxes and possibly associated with trying to explain to his wife how he lost $100 gambling or with a very small probability of an enormous revenue.

This same example, which suggests so vividly how any proper scoring rule might be applied to elicit from a geologist his opinions about the yield of a well that is to be drilled also serves to illustrate what seems to be an insurmountable obstacle to the application of arbitrary scoring rules, and even of any proper scoring rules to all, to many situations in which opinion is sought as a basis for decision. If whether the well is to be drilled does not depend on this geologist's opinion, then any proper scoring rule concerning its expected yield can be implemented, in particular by offering to sell shares in the well at various prices. But if whether to drill the well depends in part on the opinion of the geologist, then some of the events about which his opinion is wanted may never be tested precisely because of his advice, so the scoring-rule contract cannot be offered literally. The phenomenon is ubiquitous. We can never know what would have happened had surgery been ventured, had a certain product been marketed, or had a certain student been admitted. Business sharing, to be discussed in Subsection 10.12, seems to offer some possibility of circumventing this widespread obstacle to the literal application of proper scoring rules.

10.4 Big Money

The interrogator has an interest in making the possible fluctuations in the wealth of the expert large. For this motivates the expert to reflect hard and well before answering and yet need not add systematically to his expected fee. The larger the fluctuations are, however, the more the expert is motivated to report not his real probabilities but numbers that reflect in part the nonlinearity of his utility. The theory of these distortions has been somewhat explored by Winkler and Murphy [60], and some possible ways to avoid them or compensate for them were reported on in Section 3. My own hope and expectation is that the skillful use of small, or even purely symbolic, scoring-rule payments (in addition to the usual compensations) will enable experts to know and to communicate their opinions much more accurately than has been usual.

The only practical experience with elicitation involving substantial cash prizes and losses thus far apparently consists of observations on gambling behavior that Ward Edwards has made, and expects soon to publish. The use of substantial, or even of token, scoring-rule payments in business would often be a considerable break with tradition likely to involve legal and other administrative problems.

10.5   Tiny Probabilities and the Expert

Probabilities corresponding to odds of one in a million or even less, such as the probabilities of specific disasters are sometimes important [32]. Any scheme to give an expert a serious cash incentive to reveal his personal probabilities for very improbable events would seem to court insuperable difficulties with the nonlinearity of utility. And, even if we do not despair of obtaining sincere opinions about such matters by engaging sincerely inclined experts in the right kind of make-believe, the difficulty of such make-believe, and the special training required for it, are in danger of being greatly underrated.

Magnification is sometimes possible. For example, though it would not be practical to engage me in a meaningful bet that the birth awaited by the recently married Smiths will be quadruplets, my expectation for the number of sets of quadruplets among the next million American births (about two, based on a little reading) could be elicited and might be of some use. But this relief is largely illusory. For a real expert on multiple births would be expected to take into account such data as that Mrs. Smith is herself a twin and a very young bride, so there is no practical possibility of counting the quadruplet births in a large number of cases similar in the respects considered pertinent.

10.6   Employing Expert Opinion

In just what ways can you expect to profit from the opinons of experts in serious matters? According to a very broad model, you have an important decision to make in the light of all sorts of data at your disposal, and this data may include the behavior of experts. You could, in principle, explore all sorts of ways of interrogating an expert—not confining yourself to eliciting his opinions—and study empirically how his responses can profit your business. Conceivably, pain in the weatherman's great toe would better help you plan picnics than would his opinion about the weather. Yet I presume that ordinarily little of importance would be lost if you could obtain only the opinion of the expert, that is, his personal probabilities. What should be done with such an opinion? The simplest thing, and sometimes the appropriate thing, would be to make the expert's opinion your own.

This is by no means mandatory. For example, an expert who always ascribes very small probability to what actually occurs and to nothing else would be as useful as one who is omniscient, but you would of course not make the opinions of this perfect fool your own. Again, you might discover with experience that your expert is optimistic or pessimistic in some respect and therefore temper his judgments. Should he suspect you of this, however, you and he may well be on the escalator to perdition.

10.7 Divergent Opinion

We often have access to more than one expert, and what to do when doctors disagree has always been, and will always be, a quandary. One important thing to do, but far outside the scope of this paper, is to encourage the right kind of communication between the experts. Exploration of how to do this is, for example, one of the aims of the Delphi technique [3; 8; 29, Part II]. In general, good communication is what makes the experts share factual information and help each other think their opinions through thoroughly, and bad communication is what encourages various vices such as exaggeration and excessive deference. Sooner or later, despite all techniques of communication, divergent expert opinions will have to be faced. Perhaps you will make some composite of one or more expert opinions and your own opinion. An extreme way to do that would be to decide, on the basis of past experience or otherwise, that a particular one of the experts is the only one worth listening to and to make his opinion your own. A more general procedure would be to average the opinions, that is, to average the probability distributions associated with the experts (possibly including yourself), giving each the weight you think appropriate. Thus, rather than simply choose one expert among several, you can choose among the infinite number of synthetic experts that constitute the convex closure of the several. (A few key references bearing on the subject of this paragraph are [10; 11; 47, Ch. 10; 52, p. 65ff; 56].)

10.8 The Expert as an Instrument

When is one expert, real or synthetic, to be preferred to another? An "expert" in this context is a mechanism, possibly with human components, generating numbers that you contemplate using instead of your own personal probabilities (or, more generally, expectations) in certain contexts. One crude, practical answer sometimes available and appropriate, is this: Employ, until you have further experience, that expert whose past opinions, applied to your affairs, would have yielded you the largest average income.

No rule of this sort can claim absolute or objective validity, and this one has been couched especially roughly for the sake of simplicity. For example, actual past experience with the experts may be extensive, moderate, meager, or absolutely lacking. When past experience is extensive, but not too extensive, the rule often has much to recommend it; when direct past experience with the experts is meager, the rule is silly; and when such experience is altogether lacking, the rule results in a tie and is therefore empty. Actually, if you have little or no past experience with the experts, you will have to ponder them in terms of whatever information it was that brought you to regard them as promising in the first place; this well finder is regarded by the whole neighborhood as infallible with the hazel fork; that one is a professor of geology and the author of an important treatise on subsurface hydrology but has never before tried to help anyone locate a well. In such a context, the subjective aspect of your decision is thrown into prominence, but no matter how much direct past experience you may have with the experts, the ultimate subjectivity of your choice among them never disappears, though its effects may become less agonizing—according to the personalistic Bayesian theory of statistics, as in [48].

When your past experience with the experts is very extensive indeed, it may become profitable for you to refine the original rough rule by dividing up circumstances into categories and confiding in different experts for different categories of decisions. Such discussion could continue indefinitely; for the situations are innumerable and tend to parallel the whole field of decision-theoretic statistics.

An interesting process for coordinating the efforts of experts in different fields known as PIP (for probability information processing) has been vigorously pursued [20].

10.9 Slippery Utilities

A different kind of complication in applying, or adapting, the rough rule is that the notion of average income may not be readily applicable to your affairs. For one thing, it may be important to measure your own income in utility rather than in cash. If this involves only determining your utility for cash, you may be able to do that reasonably well with moderate effort. If, however, the consequences of your act are not easily converted into cash but involve values difficult to weigh against each other such as beauty, justice, and health, your dilemma may be especially severe. Raiffa [44] has recently published a book largely on these subtle problems of pondering the imponderable and evaluating the invaluable that reviews, and contributes to, a considerable literature.

Another difficulty in measuring utility is seen in this example: You are the person responsible for choosing which of several televised meteorological forecasters shall serve your city. There is, I assume for simplicity, abundant evidence of past performance, and the members of the community who use the forecasts will behave in accordance with the probabilities announced in them. Since yours is a public trust, you would like to choose the forecaster that in the past would have maximized the mean income of members of the community.

Subtle welfare-economic decisions about the relative importance of bent-pin anglers and barn painters could complicate your problem, but, even more important, relatively little is really known about the uses to which public weather information is put and what its economic consequences are. Thought has been given to the problem of the economic value of meteorological forecasts, both for the general public and for special purposes, but difficult, important empirical aspects of the question remain to be explored. (Key references are [36, 40].)

10.10   Which Scoring Rule for the Trained Respondent?

There are as many proper scoring rules for a trichotomy, for example, as there are convex functions over the baricentric triangle, or two dimensional simplex. It would therefore seem important to study in what respects one scoring rule is better than another. But this question has thus far proved surprisingly unproductive. Its elusiveness is brought out by the consideration than an ideal subject responds to all proper scoring rules, including those involving extremely small payments, in exactly the same way. Therefore, any criteria for distinguishing among scoring rules must arise out of departures of actual subjects from the ideal.

Since we all do depart markedly from the ideal, it might seem that one proper scoring rule would be much more effective with a real person than another, and this presumably is often so. But if a person is reasonably sophisticated, though far from ideal, the form of a proper scoring rule for eliciting his probabilities, for, say, a trichotomy should—provided its amplitude is sufficient to command attention—have little or no effect on his response. To see this, put yourself in his place. You are offered a contract that will result in certain cash payments to you depending on your choice of three numbers p, q, and r and on whether a certain game ends in win, lose, or tie. If you know what personal probabilities are and understand that the

the contract is so drawn that it is to your interest to report your personal probabilities, then the details of the contract seem unimportant; for no matter which proper scoring rule it corresponds to, you should ask yourself what your personal probabilities for the three events are and report them.

Yet, the terms of the contract might make a modest difference to you. Suppose, for example, that very little money is to change hands in case of a tie, no matter what your response is. In this case, you have little incentive to ask yourself carefully the probability of a tie and are thus left free to focus on the relative probability of a win given that there is not a tie. In this case, your questioner will be well served if he is mainly interested in that conditional probability, and he will be badly served if he particularly wants to know the probability that you attach to a tie.

Thus, at least a vague criterion applicable even to sophisticated but human respondents emerges. Insofar as responding is hard work, the scoring rule should encourage the respondent to work hardest at what the questioner most wants to know. If this is to be effective, it must not merely be mathematically true but also plain to the respondent that he will be rewarded most for working on the right aspects of his opinion. This is in part a psychological question of human communication, subject to much speculation and experimentation. To illustrate, you are faintly curious to know whether the respondent thinks that a tie is likely and desperate to know whether he believes that the home team will win. You might be best served by entering into two palpably separate contracts with the respondent, a small one hinging only on whether there is a tie and resulting in a rough casual elicitation of the respondent's r, and a larger one, involving no payment in case of a tie, resulting in a well considered evaluation of the ratio $p:q$ for the respondent. Of course the two contracts together amount to a single scoring rule, though presenting them separately might work better psychologically.

The appropriate incentive for you to offer a respondent for his opinion depends not only on the importance for you of obtaining that opinion with a specified accuracy but also on the difficulty for the respondent in obtaining it from himself. This makes the choice of a scoring rule designed to evoke the right degree of effort from the respondent on the various components of his task particularly subtle.

10.11   There Are No Bargain Scoring Rules

Since a scoring rule is a scheme of payments, it might seem natural to choose the rule that promises to obtain the required information as cheaply as possible or perhaps the one that obtains the most for the money. There may be something to this line of thinking, but not in first approximation, and I have been unable to make progress with it.

The respondent will presumably work for what he regards as an acceptable wage. In the presence of a scoring rule, he will perceive his wage as random with an expectation that can be adjusted by adding a constant to the scoring rule. So, within the linearity approximation, the respondent does not charge extra for submitting to the scoring rule, and it therefore seems roughly reasonable to reckon that the questioner is not charged for it.

The higher the amplitude of the scoring rule, the more incentive it gives the respondent to reply with care. On this account, the respondent might in principle come to insist on a higher mean wage for facing a highly fluctuating scoring rule, and this could tend to deter the questioner from using high amplitudes. But the important practical limitation on amplitude would seem to be the need to avoid the distortion of response induced by the nonlinearity of the respondent's utility.

10.12 Business Sharing

In common sense, we feel without any overt reference to economic models that some responses are not so wrong as others and ought not to be so heavily penalized. If a respondent is pretty sure that the home team will win, and there is in fact a tie, then he is perhaps not so wrong as if it had lost. Scoring rules reflecting this idea have been sought and easily found. See, for example, [21, 53]. One interesting way to adjust the rewards and penalties of the respondent to the interests of the interrogator, which was brought out in a dramatic and more radical form in McCarthy's [37] pioneering note on proper scoring rules, is to give the respondent a fractional interest in the business involved. To illustrate with an overidealized example, an oil prospecting company could give its geologist a small fraction of all profits and losses with the understanding that all decisions in the business would be made using the geologist's personal probabilities about geological uncertainties.

(It would be interesting to consider with some care the respects in which such an example is realistic and unrealistic, but I can only go a step or two in that direction here. Stock in the company would seem to give the geologist an interest in reporting his probabilities honestly if he could be assured that they would be adopted as the personal probabilities of the management for the events concerned. But this incentive may not be fully in harmony with the expert's incentives to appear worthy, as opposed to simply being worthy, of retention and promotion—a complication that affects not only business sharing but all applications of proper scoring rules to a professional expert. Business sharing does not present the expert with an explicit scoring rule in any business complicated enough to provide a more than mechanical role for managers, in particular in any business in which there are other uncertainties than those about which the expert is consulted, but an implicit rule is as effective in principle as an explicit one.)

Long ago, Gauss [25, Sec. 6] proposed that economic losses (such as those in a game of chance) provide a good model for the incentive to estimate accurately even in the most academic contexts. Decisively to uphold or to overthrow this suggestion does not seem possible. Personally, it appeals to me. Correspondingly, when we say that a tie should not be regarded as so distant from a win as a loss would be, I am inclined to think that that is because we have in mind various uses for sport forecasts in which the penalty is less for one kind of error than another. Of course, the penalty need not be a monetary one; it might involve, for example, loss of social prestige. Fortunately, the elusive question of whether all that is good and bad about a forecast can ultimately be referred to profit and loss in economic decisions, sufficiently, widely interpreted, need not be resolved in order to show the interest and utility of viewing proper scoring rules as a share in a real or a fictional business. For the business model is certainly a mathematically general model for all proper scoring rules and a fertile point of view for the generation of proper scoring rules that penalize some errors more than others.

The technical point that every proper scoring rule can be viewed as a share in a business and that every such share leads to an at least weakly proper scoring rule should be appreciated. Section 7 makes these points clear. Every strictly proper scoring rule amounts to the possibility of choosing among acts, only one of which is appropriate to each system of personal probabilities. Conversely, a person knowing that an act in a specified economic situation is to be chosen for him in accordance with his announced system of personal probabilities will have no incentive to announce a false one. However, if the convex function arising from the family of acts has flat places—technically, is not strictly convex—then the person has no positive incentive to distinguish among certain systems of probabilities, and the scoring rule is only weakly proper.

Insofar as business sharing is a practical method of elicitation, it makes possible the use of a proper scoring rule even in those situations stressed at the end of Subsection 10.3 in which arbitrary proper scoring rules cannot be implemented because some of the conditional probabilities to be elicited will not be tested, depending on the opinions expressed by the expert.

The parallelism between implementing a scoring rule by business sharing and the rough rule for rating experts according to how their advice would have affected your business in the past (discussed in Subsection 10.8) is evident, but the two things must not be confused. In particular, the rating rule can be used regardless of what scoring rule if any is used to elicit the opinions.

10.13   Some Armchair Psychology

Consider now a subject quite untrained in personal probability and the theory of scoring rules whose only incentives are provided by the scoring itself. It is questionable whether, in any serious application, this ought to be allowed to happen. Though any strictly proper scoring rule is a sufficient guide for an ideally intelligent Robinson Crusoe no matter how uninstructed, can we expect real people to respond well with no other coaching than is provided by the scoring rule itself or even by extended experience with the scoring rule? And even if investigation should yield a somewhat affirmative reply, is there any point in withholding instruction? Whatever the answers, it does seem stimulating to speculate on what kinds of scoring rules presented in what way would most nearly operate on naive subjects as all strictly proper scoring rules are supposed to operate on sophisticated ones, and this should lead to ideas of practical value for subjects of intermediate sophistication.

In the first place, the subject must understand the scoring rule. If it makes explicit reference to logarithms, or even to squares, most ordinary people will not understand it at all; and even those with mathematical training may not be nearly apt enough at calculation to use the rule effectively. This is an important reason to present the rule through some vivid tabular or graphic device, which could, for example, take the elaborate form of conversational-mode digital computation, or more simply of some slide-rule device such as those of the Shuford-Massengill Corporation (Lexington, Mass.), or perhaps tabulation of the scoring rule itself, possibly very boldly rounded [14].

Perhaps it is helpful to a subject responding about a partition of possible events if the economic consequence of his response is a function only of the element of the partition that happens to obtain. This condition imposes no constraint at all for two-fold partitions, but for n-fold partitions with $n > 2$, it leads to the logarithmic scoring rules of Subsection 9.4. The possible advantages of the simplicity might often be outweighed by the inappropriateness of a symmetric scoring rule in asymmetric situations or of a scoring rule that lays emphasis on the correct elicitation of small probabilities. I have sometimes heard the possibility that a subject responding to a logarithmic scoring rule could be subjected to an infinite (or at any rate, unlimited) penalty raised as an overwhelming objection. This possibility does of course imply that the method cannot be applied literally, but approximate applications, in which the subject is not allowed to name probabilities less than, say, 10^{-3} suggest themselves. And, as mentioned in Subsection 10.5, obtaining probabilities very close to 0 by means of direct incentives does not seem practical by any scoring rule.

10.14　Proper Scoring Rules in School

Proper scoring rules hold forth promise as more sophisticated ways of administering multiple-choice tests in certain educational situations [14, 49]. The student is invited not merely to choose one item (or possibly none) but to show in some way how his opinion is distributed over the items, subject to a proper scoring rule or a rough facsimile thereof.

Though requiring more student time per item, these methods should result in more discrimination per item than ordinary multiple-choice tests, with a possible net gain. Also they seem to open a wealth of opportunities for the educational experimenter.

Above all, the educational advantage of training people—possibly beginning in early childhood—to assay the strengths of their own opinions and to meet risk with judgment seems inestimable. The usual tests and the language habits of our culture tend to promote confusion between certainty and belief. They encourage both the vice of acting and speaking as though we were certain when we are only fairly sure and that of acting and speaking as though the opinions we do have were worthless when they are not very strong.

Effects of nonlinearity in educational testing deserve some thought, but presumably nonlinearity is not a severe threat when a test consists of a large number of items. One source of nonlinearity that has been pointed out to me is this. A student competing with others for a single prize is motivated to respond so as to maximize the probability that his score will be the highest of all. This need not be consistent with maximizing his expected score, and presumably situations could be devised in which the difference would be important.

REFERENCES

[1]　Aczel, J., *Lectures on Functional Equations and Their Applications,* New York: Academic Press, 1966.

[2]　Allais, Maurice, "Le Comportement de l'Homme Rationnel Devant le Risque: Critique des Postulats et Axioms de l'École Americaine, *Econometrica,* 21 (October 1953), 503-546.

[3]　Ament, Robert H., "Comparison of Delphi Forecasting Studies in 1964 and 1969," *Futures,* 2 (March 1970), 35-44.

[4]　Bayes, Thomas, "Essay Towards Solving a Problem in the Doctrine of Chances," *The Philosophical Transactions,* 53 (1763), 370-418. Reprinted in W. Edwards Deming, ed., *Facsimiles of Two Papers by Bayes,* Washington, D.C.: The Graduate School, The Department of Agriculture, 1940, and in "Thomas Bayes's Essay towards Solving a Problem in the Doctrine of Chances," *Biometrika,* 45 (December 1958), 293-315.

[5]　Brier, G. W., "Verification of Forecasts Expressed in Terms of Probability," *Monthly Weather Review,* 78 (January 1950), 1-3.

[6] Brown, T. A., "Probabilistic Forecasts and Reproducing Scoring Systems," Santa Monica, Calif.: RAND Corporation, RM-6299-ARPA, June 1970, 65 pp.

[7] Buehler, Robert J., "Measuring Information and Uncertainty," in *Proceedings of the Symposium on Foundations of Statistical Inference*, Toronto: Holt, Rinehart, and Winston, 1971 (in press).

[8] Dalkey, N. C., "An Experimental Study of Group Opinion: The Delphi Method," *Futures*, 1, 5 (September 1969), 408-26.

[9] de Finetti, Bruno, "La prévision: Ses Lois Logiques, Ses Sources Subjectives," *Annales de l'Institute Henri Poincaré*, 7 (1937), 1-68. English translation in (Kyburg and Smokler, 1964).

[10] —————, "La Notion de Distribution d'Opinion comme Base d'un Essai d'Interpretation de la Statistique," *Publications de l'Institut de Statistique de l'Université de Paris*, 1, 2 (1952), 1-19.

[11] ————— , "Media di Decisioni e Media di Opinioni," *Bulletin de l'Institut International de Statistique*, 28th session, 34, 2 (1954), 144-57.

[12] ————— , "Does It Make Sense to Speak of 'Good Probability Appraiser'?" in *The Scientist Speculates: An Anthology of Partly-Baked Ideas*, New York: Basic Books, 1962, 357-64.

[13] ————— , "Probabilità Composte e Teoria Delle Decision," *Rendiconti de Matematica*, 23 (1964), 128-34.

[14] ————— , "Methods for Discriminating Levels of Partial Knowledge Concerning a Test Item," *The British Journal of Mathematical and Statistical Psychology*, 18 (May 1965), 87-123.

[15] ————— , "Quelques Conventions Qui Semble d'Être Utile," *Revue Roumaine de Mathematique Pures et Appliquées*, 12, 9 (1967), 1227-33.

[16] ————— , "Probability: Interpretations," Vol. 12, 496-505, in the *International Encyclopedia of the Social Sciences*, New York: The Macmillan Company, 1968.

[17] ————— , "Logical Foundations and Measurement of Subjective Probability," *Acta Psychologica*, 34 (December 1970), 129-45.

[18] ————— , and Leonard J. Savage, "Sul Modo di Scegliere le Probabilità Iniziali," *Sui fondamenti della statistica, Biblioteca del Metron*, Series C, 1 (1962), 81-154 (English summary, 148-151).

[19] Edwards, Ward, "A Bibliography of Research on Behavioral Decision Processes to 1968," *Memorandum Report No. 7*, Human Performance Center, University of Michigan, January 1969, 93.

[20] ————— , L. D. Phillips, W. L. Hays, and B. C. Goodman, "Probability Information Processing Systems: Design and Evaluation," *IEEE Transactions on Systems Science and Cybernetics*, SSC-4 (1968), 248-65.

[21] Epstein, Edward S., "A Scoring System for Probability Forecasts of Ranked Categories," *Journal of Applied Meteorology*, 8 (December 1969), 985-7.

[22] Fishburn, Peter C., *Utility Theory for Decision Making*, New York: John Wiley and Sons, Inc., 1970.

[23] Friedman, Milton, "Choice, Chance, and Personal Distribution of Income," *Journal of Political Economy*, 61, 4 (August 1953), 277-90.

[24] ————— , and Leonard J. Savage, "The Utility Analysis of Choices Involving Risk" *Journal of Political Economy*, 56 (August 1948), 279-304. Reprinted in *Readings in Price Theory*, Chicago: Richard D. Irwin, 1952, and in *Landmarks in Political Economy*, edited by Earl J. Hamilton, *et al.*, The University of Chicago Press, 1962.

[25] Gauss, Carl Friedrich, "Theoria Combinationis Observationum Erroribus Minimis Obnoxiae," *Commentationes Societatis Regiae Scientarum Gottingenis Recentiores,* 5 (1821), 33-90. In German translation, A. Börsch and P. Simon, *Abhandlungen zur Methode der kleinsten Quadrate,* Berlin: Westdruckerei Jaochim Frickert, 1887 (reprinted Wurzburg: Physica-Verlag, 1964). In French translation, J. Bertrand, *Methode des Moindres Currés,* Paris: Maalet-Bachelier, 1855.

[26] Good, I. J., "Rational Decisions," *Journal of the Royal Statistical Society, Ser. B,* 14 (1952), 107-14.

[27] Grayson, Charles Jackson, *Decisions under Uncertainty: Drilling Decisions by Oil and Gas Operators,* Boston: Harvard University, Division of Research, Graduate School of Business Administration, 1960.

[28] Haavelmo, Trygve, "The Role of the Econometrician in the Advancement of Economic Theory," *Econometrica,* 26 (July 1958), 351-7 (Presidential address, Econometric Society, Philadelphia, Dec. 29, 1957).

[29] Helmer, Olaf, *Social Technology,* New York: Basic Books, 1966.

[30] Hendrickson, Arlo D., "Payoffs to Probability Forecasters," *Technical Report No. 145,* University of Minnesota, September 1970.

[31] Kyburg, Henry E., Jr., and Smokler, Howard E., *Studies in Subjective Probability,* New York: John Wiley and Sons, 1964.

[32] Maloney, Clifford J., "A Probability Approach to Catastrophic Threat," *ARO–D Report 70-2,* U.S. Army Research Office–Durham, Durham, North Carolina, 1970?.

[33] Marschak, Jacob, "Management Research and Behavioral Science," *Working Paper No. 2,* Western Management Science Institute, University of California, Los Angeles, 1961.

[34] —————— , "Actual Versus Consistent Decision Behavior," *Behavioral Science,* 9 (April 1964), 103-10.

[35] —————— , "Decision Making: Economic Aspects," Vol. 4, 42-5, in the *International Encyclopedia of the Social Sciences,* New York: The Macmillan Company, 1968.

[36] Maunder, W. J., *The Value of the Weather,* London: Methuen, 1970.

[37] McCarthy, John, "Measures of the Value of Information," *Proceedings of the National Academy of Sciences,* (1956), 654-5.

[38] Mosteller, Frederick, and Wallace, David L., *Inference and Disputed Authorship: The Federalist,* Reading, Mass.: Addison-Wesley, 1964.

[39] Murphy, Allan H., "The Ranked Probability Score and the Probability Score: A Comparison," *Monthly Weather Review,* 98 (December 1970), 917-24.

[40] —————— , and Allen, R. A., "Probabilistic Prediction in Meteorology: A Bibliography," Silver Spring, Md., Environmental Science Services Administration, Technical Memorandum No. WBTM TDL 35, 1970.

[41] —————— , and Epstein, Edward S., "Verification of Probabilistic Predictions: A Brief Review," *Journal of Applied Meteorology,* 6 (October 1967), 748-55.

[42] —————— , and Winkler, Robert L., "Scoring Rules in Probability Assessment and Evaluation," *Acta Psychologica,* 34 (December 1970), 273-86.

[43] Pratt, John W., "Risk Aversion in the Small and in the Large," *Econometrica,* 32, 1-2 (January-April 1964) 122-36.

[44] Raiffa, Howard, *Decision Analysis: Introductory Lectures on Choices under Uncertainty,* Reading, Massachusetts: Addison-Wesley, 1968.

[45] Roberts, Harry V., "Risk, Ambiguity, and the Savage Axioms: Comment," *Quarterly Journal of Economics,* 77 (1963) 327-42.

[46] _____ , "On the Meaning of the Probability of Rain," in *Proceedings of the First National Conference on Statistical Meteorology,* Boston: American Meteorological Society (1968) 133-41.

[47] Savage, Leonard J., *The Foundations of Statistics,* New York: John Wiley & Sons, Inc; and London: Chapman and Hall, 1954.

[48] _____ "The Foundations of Statistics Reconsidered," in *Proceedings of the Fourth Berkeley Symposium,* Berkeley, University of California Press (1961) 575-86. Reprinted in (Kyburg and Smokler, 1964).

[49] Shuford, Emir H., Jr., Albert Arthur, and U. Edward Massengill, "Admissible Probability: Measurement Procedures," *Psychometrika,* 31 (June 1966), 125-45.

[50] Smith, Cedric A. B., "Consistency in Statistical Inference and Decision," *Journal of the Royal Statistical Society,* Ser. B., 23 (1961), 1-25.

[51] _____ , "Personal Probability and Statistical Analysis," *Journal of the Royal Statistical Society,* Ser. A., 128 (1965), 468-89.

[52] Staël von Holstein, Carl-Axel S., *Assessment and Evaluation of Subjective Probability Distributions,* Stockholm: The Economic Research Institute at the Stockholm School of Economics, 1970.

[53] _____ , "A Family of Strictly Proper Scoring Rules Which Are Sensitive to Distance," *Journal of Applied Meteorology,* 9 (1970), 360-4.

[54] Toda, Masanao, "Measurement of Subjective Probability Distribution," *Report No. 3,* Division of Mathematical Psychology, Institute for Research, State College, Pennsylvania, 1963.

[55] Winkler, Robert L., "The Assessment of Prior Distributions in Bayesian Analysis," *Journal of the American Statistical Association,* 62 (September 1967), 776-800.

[56] _____ , "The Consensus of Subjective Probability Distributions," *Management Science,* 15, 2 (October 1968), B-61-B-75.

[57] _____ , "Scoring Rules and the Evaluation of Probability Assessors," *Journal of the American Statistical Association,* 64 (September 1969), 1073-78.

[58] _____ , and Allan H. Murphy, "Evaluation of Subjective Precipitation Probability Forecasts," in *Proceedings of the First National Conference on Statistical Meteorology,* Boston: American Meteorological Society, 1968, 148-57.

[59] _____ , and Allan H. Murphy, " 'Good' Probability Assessors," *Journal of Applied Meteorology,* 7 (October 1968), 751-8.

[60] _____ , and Allan H. Murphy, "Nonlinear Utility and the Probability Score," *Journal of Applied Meteorology,* 9 (February 1970), 143-8.

BRUNO DE FINETTI

Probability: Beware of Falsifications!
(1977)

Reprinted from *Scientia*, 111, 1976, pp. 283-303 with permission of the
editor and the author.
[Certain misprints and errors in translation have been corrected.]

INTRODUCTION - I

On this occasion I should like to clarify in the simplest, but at the same time most complete and radical way, the criticism of the large numbers of ways in which (to my mind) probability is misunderstood and misrepresented. The problems and aspects involved are numerous, complex and intricate. Nevertheless, even if a speech is not enough to illustrate a certain conception and to convince people to adopt it, it can at least clarify any doubts raised against it and raise doubts concerning the opposite conceptions.

Basically, I have never found anything to modify in, or to add to, (although it has been extended and deepened) the conception I gradually worked out between 1926 and 1928, the years immediately before and after my graduation (Univ. of Milan, 1927). It was in 1926 that a popularizing article by Carlo Foà on Mendel's laws of heredity made me stop and think about probability and read Czuber's *Wahrscheinlichkeitslehre*, a copy of which was lent to me by an acquaintance, and it was in 1928 that—in a communication read at the Lincei (published there in 1931) and summarized in a communication to the International Congress of Mathematics held at Bologna—that I introduced the concept of 'exchangeability'.

The sensational effect of this concept (which went well beyond its intrinsic meaning) is described as follows in Kyburg and Smokler's preface to the collection *Studies in subjective probability*, which they edited (p. 15 in this edition): "In some ways the most important concept of the subjectivistic theory is that of *exchangeable events*. Until this notion was introduced by de Finetti in 1931, the subjectivistic theory of probability remained pretty much of a philosophical curiosity. None of those for whom probability theory was a means of livelihood or knowledge paid much attention to it. But with the introduction of the concept of "equivalence," or "symmetry," or "exchangeability," as it is now known, a way was discovered to connect the notion of subjective probability with the classical procedures of statistical inference."

To be precise, the apt term 'exchangeability' was suggested later by Fréchet (1939) who spoke of *événements échangeables*. This term was rapidly adopted by a number of authors, including myself (and it became *exchangeable* in English and *austauschbar* in German). In my early work I stated the above condition without introducing any denomination. Between 1933 and 1939 I used the term *equivalent* as proposed by Khinchin in 1932 (and still in use in the USSR); other (less common), denominations are *symmetrical* or *invariant under mixing* (in fact we are dealing here with symmetry or invariance, with respect to permutations[1]). I attempted a first synthetic but global exposition of my point of view at the request of the Institut Poincaré, which invited me to give a series of lectures (2-10 May 1935). The French text is now hard to find, but the excellent English translation by Leonard Jimmie Savage* is to be found in the slim volume by Kyburg and Smokler mentioned above (first edition, Wiley, 1964), together with articles by John Venn, Emile Borel, Frank Plumpton Ramsey, Bernard O. Koopman and Savage himself.

I cannot help taking the opportunity of recalling this figure who was so exceptionally rich in critical spirit, extensive knowledge (in all fields) and extraordinary humanity. In particular I must stress that I owe it to him if my work is no longer considered as a blasphemous but harmless heresy, but as a heresy with which the official statistical church is being compelled, unsuccessfully, to come to terms. Savage's cooperation has been of enormous help to me, not only because of his extraordinary and many-faceted qualities but also because his doubts on the dogmas of the 'objectivistic' official statistical church came to maturity after he had been raised in her, in the form of a painful impression in someone who, in the face of absurdities and imperfection, feels the faith in which he had been raised to believe melting away. I, of course, was in quite the opposite situation—a layman or barbarian who had the impression that the others were talking nonsense but who virtually does not know where to begin as he cannot imagine the meaning that the others are trying to give to the absurdities that it is imposing as dogma.

On the subject of acknowledgments, I should of course have to mention many other names; for obvious reasons, however, I shall limit myself to those three colleagues who, although not sharing my opinions (but their merit is all the greater because of this), have helped me and given me the opportunity of presenting my ideas in the right quarters, that is, Guido Castelnuovo, Maurice Fréchet and Jerzy Neyman.

After that of the Institut Poincaré and many others of the same kind, the present exposition is intended to be a new and final attempt to coordinate

(1)   The work in which Rene Maurice Fréchet introduced the term *échangeable* is *Les probabilités associées à un système d'événements compatibles et dépendants*, Hermann ed., Paris, 1939. As far as I know the first to use it in English were Chernoff and Teicher (1957), and in German Buehlmann (1960).

       As for the other terms: *equivalent* was introduced by A. Khinchin in *Sur les classes d'événements équivalents*, Mat. Sbornik (1932), followed by Milicer-Gruzewska (1950), Dynkin (1953), Ryll-Nardzewski (1957) and still (it seems) in USSR; that of *symmetric* was used by E. S. Andersen in *On sums of symmetrically dependent random variables*, Skandin. Aktaurietidskrift (1953) and (seq.) (1954), and by Hewitt and Savage (1955); the more explicit but longer expression *invariant under mixing* by D. Freedman (1962).

       *The translation is in fact by Kyburg.

in a synthesis that is as systematic as possible (but not too specialistic, almost 'without formulae, theorems and the like') considerations of a logical, mathematical, psychological, economic, epistemological or (if you prefer) philosophical nature—aimed at clarifying which formulations, approaches and conclusions of a conceptual and mathematical nature are or not admissible and reasonable—*and why*—in the context of a global consistent conception.

Basically, I shall be repeating, for the most part, things that have already been said many times, but many of the observations and arguments (as well as their coordination) are new or renewed.

PERIPATETICS OR PRAGMATISTS? - II

The fact is that, in my opinion, in the conceptions in vogue the notion of probability and its properties (when they are not sterilized—and that is worse still!—in empty, formalistically abstract axiomatic constructions and, as is perhaps inevitable, constructions that are basically imperfect) are linked to reality and to applications by means of circular or senseless assertions. I should not like to seem too harsh, but I cannot help confessing that I find an extraordinary analogy between the type of contortions that are repeated in defense of such assertions and the way of reasoning at which the legendary Simplicius, immortalized in Galileo's *Dialogues,* was so clever. But, unfortunately, that is how matters stand whether we like it or not.

I am quite prepared to apologize for my insistence, but it does not seem to be excessive (or even enough, for that matter): it is due to my anxiety over the danger that seems to be hanging over studies on probability and its applications, whose peculiar and consistently unitary character seems to be threatened by the proliferation of confused ideas, partial views and artificial techniques. (e.g. The rules of thumb and tests to describe which Irving Good has coined the apt term of "adhockeries"). For reasoning to be valid it must follow a diametrically opposite course: it is necessary to follow the example very appropriately described in a phrase that—ever since I read it as a boy—has remained impressed as a "memento!," as a rule to be followed and to be recommended to all. Here it is:

"His concern was to teach the prudence and tricks by means of which it is possible to succeed in formulating propositions that have a meaning."

This quotation was taken from Giovanni Papini (*Stroncature,* n. 14), it refers to Mario Calderoni who—together with Giovanni Vailati—was one of the few courageous supporters of pragmatism in Italy at the end of the 19th and beginning of the 20th centuries.

To formulate *propositions "that have a meaning,"* making sure one does not fall into the 'hundred ways of not saying anything' so dear to rhetoricians

and cheats,[2] is a particularly difficult task—as Calderoni well knew—but
also *unusual* and above all *irritating* for the would-be right-minded people
who, in their immense slovenliness, claim that all their ravings have a meaning
(and often, in reciprocation, they admit that also someone else's ravings
have a meaning too).

Also contributing to the confusion of tongues and ideas is that 'tyranny
of language' denounced by Harold Jeffreys in his acute observation that
"language was created by realists, and very primitive realists at that" so that
*"we have a very large number of possibilities of describing the properties
attributed to objects, but very few of describing those directly known as
sensations."*[3]

And so also the notion of probability, instead of being considered and
studied as such by trying to perfect our understanding of it and its use, is
often exteriorized, as it is believed to be conceivable only if represented—if
not like a real 'object'—at least like 'something' (goodness knows what?!)
that exists outside ourselves, something that 'acts' on the outside world
according to its own 'laws' which are supposed to govern the facts which
do not follow 'laws' (in the strict, deterministic, sense) with a surrogate of
'laws' that are not 'laws'.

'Puzzles' such as these (obviously!) lead to an almost general incompre-
hension of probabilistic reasoning, which becomes entangled in a web of
stock phrases, hackneyed sophisms and noxious ambiguities that make it
hard to unmask it.

Moreover, that the task—of dissipating obscureness and metaphysical vapours
—in the field of probability is a particularly difficult one is amply demonstrated by the—now famous and also unchallengeable—phrase by Garrett
Birkhoff (*Lattice Theory*): *"Everyone talks of probability, but no one can
explain what he means by probability in a way that is acceptable to others"*.
Unfortunately there is something worse than the *non-knowledge* of a concept,
of the meaning of a word in a given, necessary terminology: and that is the
pseudo-knowledge of some *'near enough meaning'.* It does happen, unfortunately, that, on hearing the same word (that is more or less mysterious and in any
case new to him) repeated in the context of various sentences whose general
meaning he thinks he can grasp, a person ends up by associating it with some
'near-enough meaning' which, however, crumbles into a thousand contradictions at any attempt to define it exactly. This is the problem—I suppose—
Goethe was referring to when he wrote the two well-known verses:

(2)   *Pragmatism and the hundred ways of saying nothing* is the title of an article by
      Vailati and Calderoni that was to have been included in a book on Pragmatism,
      which was interrupted on Vailati's death (Cf. G. Vailati, *Scritti*, ed. Seeber,
      Firenze, 1911).

(3)   H. Jeffreys, *Theory of Probability*, Oxford 1939 (p. 394).

"Gewöhnlich glaubt der Mensch, wenn er nur Worte hört, es müsse sich dabei auch etwas denken lassen".

(While listening to mere words, man generally thinks that underneath them lies something that must also be thought about.) In any case, a word by itself has no 'meaning': it has or it acquires a meaning only when it is placed (like a tessera in a mosaic) in a meaningful context, in a sentence or statement about which one concretely knows the meaning of saying it is *true* or *false*.

To consider Probability (with a capital P) as a metaphysical entity that exists in abstract is like thinking that it is possible (without being Alice in Wonderland) that the cat's smile can remain and continue to visible even after the cat has disappeared. I have vehemently rejected all conceptions of probability in such an absolute sense (in the preface to the English edition of my treatise) by suggesting as a motto: "PROBABILITY DOES NOT EXIST" by which I mean that probability does not "exist" on its own, independently of the evaluations we make of it mentally or instinctively. As a result, it has no meaning to wonder "what probability *is*," we should instead meditate introspectively upon which cases and in which sense we evaluate it, reason about it, and find it a suitable tool, a precious guide, for *thinking* and *acting* in *conditions of uncertainty*.[4]

However, it is obviously necessary to illustrate the reasons for this rejection, also because it clashes with views that are widely, and also authoritatively, held.

WHOSE UNCERTAINTY AND ABOUT WHAT? - III

To introduce this debate it is necessary to go back beyond the usual controversy between the objectivistic and subjectivistic conceptions of probability and to demand how to understand and delimit the field of facts (or situations or statements or whatever one wants to call them), in which one can or must (depending on the point of view) speak of *uncertainty* (and thus of probability).

A *statement,* or *proposition,* (or *'event',* to use a term from probabilistic language) is such if it expresses something that can have only two alternatives: either *true* or *false,* or YES or NO (per se: "tertium non datur"); but it is possible not to *know* the answer (with certainty) and so, at least at the outset, there are three cases: *certain, impossible, uncertain,* according to whether we know the right answer (YES or NO) or whether we DON'T KNOW. (It

---

(4)   *La probabilità: guida nel pensare e nell'agire* is the title of a series of lectures at the Univ. of Trento (1965), publ. as *Quaderno n. 11,* ivi (1966).

could obviously also happen that as a result of erroneous information, a trick
of the memory or an error in reasoning that we are "certain" of YES when
it is NO, or vice-versa. This must be taken into account, although it lies
beyond the scope of probability theory or can be included in it only as a
marginal and degenerate case.) The intermediate case, i.e. that of *uncertainty*,
is the only one requiring—or at least allowing—considerations to be made
on probability, and in a sense *related* to the state of knowledge, too. Outside
of this, only the logic of the certain exists, but also here it is nevertheless
useful to indicate, by means of 1 and 0, the 'truth values' *true* and *false*.
As has been amply illustrated, this allows one to arithmetize—and, more
specifically, linearize—Boolean algebra. This basically means extending
and using more systematically useful conventions widely employed by
Marshall Stone. Although this is mainly of use in probability calculation
(as probability, and more in general, *'prediction'* or 'mathematical hope',
is additive) I consider that this algorithm (which does not exclude, and indeed
combines superbly with the Boolean one) ought to be known and applied
also by those interested in logic and automation, as it makes it possible to
act simultaneously on events and random numbers (or 'unknowns'). This
means operating simultaneously using the operations: + (sum); - (difference);
$\times$ (product; often understood or with $\cdot$); $\sim$ (tilde) (complement to 1), i.e.
denial); $\wedge$ (less) and $\vee$ (more). (Cf. TdP, pp. 46-50; a few examples:
$E_1 \vee E_2 = \sim (E_1 \wedge E_2) = 1 - (1 - E_1)(1 - E_2) = E_1 + E_2 - E_1 E_2$ and
similarly for 3 or more events; $P(\sim E) = P(1 - E) = 1 - P(E)$, etc.)

Let us go back to the subject of *uncertainty*. Here, however, account must
be taken of the different interpretations that can be applied to it according
to various preconceived philosophic ideas or comparatively undecipherable
attitudes.[5]

In the narrower—*personalistic*—sense uncertainty about an event ceases,
for any given individual, only when he receives certain information about
it (except in the case of denial ... and then until the right news comes);
however, it is *until that time* that, on the strength of his residual uncertainty,
moment by moment, he will attribute a certain probability to it. Indeed,
as memory is not indelible, uncertainty could raise its head again as soon as
the news *is forgotten* or has become confused or uncertain.

In an intermediate—*empiristic*—sense it could be said instead that uncertainty ceased ('objectively') the moment in which the *fact* in question occurs
in the way stated or otherwise; and it is up to that time, according to this

(5)   On these conventions and notations cf. in part. paragraphs I, 10.3 and II, 11.1-4
       in TdP (thus abbrev. here and in foll. for *Teoria delle probabilità* by BdF, ed.
       Einaudi, Torino, 1970: BdF = Bruno de Finetti). English translation (*Theory of
       probability*), Wiley, Chichester, 1976.

point of view, that anyone could attribute to it that probability which corresponds to his own judgment.

In a still less narrow—*deterministic*—sense, uncertainty would never exist or would, in any case, cease to exist right from the time in which the event is found to be 'determined' in the sense that the phenomenon on which it depends takes place according to laws that are supposedly strictly deterministic without there being any possibility of other kinds of perturbative influences (e.g. human).

ELIMINATION OF MISUNDERSTANDING - IV

In order to eliminate the doubts and misunderstandings that are continuously being generated, it would be sufficient to quote the following observation by Borel: *"You can toss a coin and guess heads or tails while the coin is still in the air and its movement is perfectly determined, and also after it has fallen, on the sole condition that one cannot see on which face it is lying".* The example, itself, is banal but the conclusions implicit in it are of decisive clarificatory importance.

Whether or not one can speak of probability in connection with a certain event does not depend on the existence of an 'objective uncertainty' (however conceived) (or "uncertainty in itself," as perhaps Kant[6] might have said) as regards its occurrence (or 'having occurred') but rather on the idea that its occurrence is connected with the intervention of the Goddess of Luck, or of a mythical Chance (perhaps the "God Chance"[7]) or, (to use an expression that is no less unhappy although fashionable) of an almost personified Nature, perhaps of a 'divine' or 'diabolical' will, or lastly, by any other kind of nineteenth century 'Scientific Determinism' or the like.

(6)   This humorous reference to hazy metaphysical concepts was suggested to me by an anecdote told to me by my colleague Gaetano Fichera. On a panel of high school examiners the Philosophy examiner often asked what the *noumen* was and was content when everyone was able to give the right answer: "it's the *thing in itself.*" Fichera ended up by asking if he thought they knew what the "*thing in itself*" was; the other asked the question and the examinee answered "*it's the noumen,*" and the philosopher turned to Fichera triumphantly: "*See. They know!*"

(7)   A witty poem entitled *O Dio Caso! (Oh, God of Chance!)* was sent to me some time back by a likable and lively man of letters from Romagna, and I published it in *Periodico di Matematiche* (1974, n. 3). Everything seemed to be correct and reasonable but, later, I realized that, like many others, he believed that Chance's function was to *prevent* improbable things (like long delays in lottery number extraction, long sequences of *heads* or *double sixes* with a coin or dice, etc.) and, evidently indignant when I persisted in my *error,* he broke off the correspondence.

But what would an acceptance of this or that opinion (or 'phrase') mean to anyone having to evaluate the probability of a fact he is interested in (which, for instance, may affect the result of any decisions he makes)? Quite apart from any superstitious attitudes and consequent consultation of fortune-tellers and sorcerers, all this could only have a subjective influence on his evaluation of the probabilities (just like any other element and feature of the personality of each of us), according to how aware he is of the fact (he may even deny it) that, however it is made, his choice is a subjective one (even if made using criteria that are apparently not subjective). But this does not change the fact that it *is* in the slightest.

Finally, it could even be admitted (if no one found it odd or futile) that the 'determinism' attributed by Borel to the spinning coin begins long before, provided one can *"measure exactly the state of Caius' muscles as he is about to toss it"*. This is actually stated by Hans Reichenbach[8] who nevertheless adds that *"we could not do it: it could only be done by Laplace's superman"*. However, with equal certainty that denial was impossible, he could have said that (*if* he 'existed') Laplace's superman would be able to know all the future exactly (*"not a leaf would stir that he had not always known about it"*). It is fortunate that probabilistic reasoning is used in cases that are much more banal and hence more serious.

The above observation by Borel already contains a much more general and radical conclusion. Not only is the supposedly deterministic or indeterministic nature of the fact under consideration—i.e. of the fact on which (for instance) we agree to bet—found to be insignificant but, by the same token, also the very distinction between past and future, between things still 'uncertain' (in the 'factual' sense of 'not having occurred yet') or 'certain' (provided the result is not known). As (if I do not confuse) Poincaré said, *"probability depends partly on our knowledge and partly on our ignorance: on what we know and on what we do not know"*. Not only is it possible to bet, provided we do not know the result (and that the competitor does not either) of tossing a coin after the coin has been thrown (or on the results of a lottery drawn 20 years ago or the result of the Roma-Lazio football match in the 1929-30 championship) but, on the same condition, on any other 'certain' thing. E.g., on the value of the 50,000th decimal figure of π (already calculated and published: see Shanks and Wrench's tables up to the 100,000th figure, in *Mathematics of Computation,* 1962) or on that of the millionth (which is still not, and perhaps will never be, known by man, although it is mathematically determinate). It could be objected that the bet is practically certain to remain undecided, but this is always on the cards, were it only due to chance events, whatever the object is.

(8)   Cf. H. Reichenbach, *I fondamenti filosofici della meccanica quantistica* (1941); It. trans. Einaudi, Torino, (1954, p. 704).

PROBABILITY: WHAT IS IT? AND OF WHAT? - V

It was inevitable that the term *probability* should be used also in the above considerations which are only preliminary to the arguments intended to define its meaning and thereby give it the right to exist in an unambiguous language.

We could not then, nor can we now, suddenly ask ourselves " 'what probability is' " at the risk of being stranded on metaphysical quicksands. What we must do is to draw out the meaning of 'probability' by examining and discussing the way in which we talk about it (and evaluate it) in contexts that are suitable and significant for us.

Instead of asking ourselves "what probability is" we shall examine the meaning implicit in the use we intend to make of it and, for this purpose, we shall ask ourselves the following three questions:

- "probability of what?", and to this we shall all answer "of an *event*"; and then
- "in what circumstances?", and here it is natural to answer "taking into account all the circumstances known to be relevant at the time"; and lastly
- "evaluated by whom?" to which the only possible answer is "by the subject considering them"; if we like, we could each reply "by me".

These are the three answers a subjectivist like me would give, but it is for this very reason that I have no inclination at all to close the matter with apodictic answers like these. Their only purpose is to indicate the path to be followed in the discussion in which the various possible positions will be compared.

The unanimous answer that probability refers to *'events'* is not very, or not at all, useful as it is due to the very differences in the various customs and interpretations applied when conceiving of and using the word 'event' that irremediable obscureness and confusion over the concept of probability inevitably arise and are perpetuated.

Leaving aside less important nuances, there are basically two main types of interpretations according to whether the term 'event' is understood in a *specific* or a *generic* sense.

In the first case, 'event' is meant to refer to a certain result in a single well-defined case; that is, an *event* is a statement such that by betting on it it can be unmistakably established whether the event is *true* or *false* (it *has* or *has not* occurred) and so *whether the bet has been won or lost*.

For instance, the following propositions *are events:*

- "A draw between Juventus and Torino next Sunday (5.12.1976)",
- "The number 13 being drawn in the Rome lottery next Saturday (4. 12.1976)".
- The death by the end of 1977 of the 49-year-old insurance policy holder Mr. X. Y.", etc.

Here all is clear and unambiguous. In the second case, however, all is ambiguous and confused. In fact, the term 'event' is used in a 'collective' sense to indicate indiscriminately any one of numerous or an infinite number of possible *'events in the first sense'* which are more or less 'analogous' and are called the *'proofs'* of a certain *'event in the second sense'* (i.e. 'event' in the *generic* sense). For instance: "A draw in a (nonspecified) match in the Italian football championship, 1976-77, First Division"; "The drawing of the number 13 in any lottery and on any date"; "The death by the end of 1977 of a 49-year-old man."[9]

BAN ON AMBIGUOUS TERMINOLOGY - VI

It is quite harmless, and may often even be useful, to introduce a collective denomination to make general reference to events that share certain descriptive features (as in the examples given above) provided great care is taken to avoid any of those nominalistic misunderstandings that demand or suggest that, merely because they are grouped under the same name, the events have more in common than they actually do: in particular, in our subject, that they have equal probability or other circumstances (e.g. '(stochastic) independence', cf. n. 8) related to it, instead of *specifying them explicitly* in each particular case.

For this reason, desiring or being obliged to express the same circumstances by means of *harmless* terms (i.e. not misleading, not involving the risk of causing the above kind of confusion) one might for instance say (as I suggest and do myself) that certain *events* (again in the sense of 'single events'!) are *'evidence of one and the same phenomenon'*.

At first sight this distinction may seem to be an attempt to reduce the question to a mere verbal distinction, but this is not so, as the difference between the two expressions lies in the implicit or explicit implications *present* or *not present* in the two phrases.

An analogy will perhaps be useful: when I say "animals of the same *species*" I must use "species" as naturalists do; when I say "of the same *group*" I can refer to the group of those living today in the Hamburg Zoo, or those that are grey, or whatever else according to the case.

(9)   For the record: the first two events (actually real) were both *false* (result of Juventus-Torino, 0-2); 13 was not drawn in Rome (it came out in Venice); the third is falsely specific as it does not say who Mr. X. Y. is. However, the choice of *examples* does not in any way mean they are to be understood as forecasts or vice versa.

The implications that are *present* or *not present* are, according to the expression used, analogous in our case. They are 'present' when we say "evidence of one and the same **event**," because, according to current usage by saying this it is generally understood that this 'evidence' must automatically and acritically be considered as *equally probable* and (worse still) often also *independent,* or better, *stochastically independent,*[10] thereby running the almost certain risk of confusing or even identifying (!) probability with frequency. One fatally becomes irretrievably lost in the maze into which one is led by distorting and banalizing in this way the extensive network of significant but delicate two-way relationships that exist between probability and frequency: one the one hand, between evaluated and predicted probabilities concerning future frequencies and, on the other, between observed frequencies and any consequent modification of probability evaluation for future trials.

When we say "evidence of one and the same phenomenon" we know we are referring merely to a few external circumstances, perhaps only to common denomination, which may make the formulation of examples more expressive and convenient, and references and practical applications more specific, but which has nothing to do with probabilistic treatment. Therefore, *no* occult implications are *present.*

The following table sets out and schematicizes the comparison between the two terminologies and clarifiies the basic differences.

Terminology	*Objectivistic*	*Subjectivistic*
name of single case	evidence of an event	event (evidence of a phenomenon)
id. in collective sense	event	phenomenon
AMBIGUITY?	YES	NO

But at this stage all that has to be done is to introduce the concept of 'exchangeability' (mentioned in the opening remarks) in order to find in an impeccably *correct form* those conclusions that the use of unappropriate

(10) It would be advisable always to say *stochastically independent* (and *stochastic independence,* etc.) except where and when one can be sure that to understand *stochastic* does not give rise to misunderstanding with *logical* or *linear* or *conditioned stochastic* independence which would often mean exchangeability; see para VIII. Furthermore, it must not be thought that independence refers to the *events* in themselves; it refers merely to our evaluation of probability, of $P(AB) = P(A)P(B)$, etc.

expressions not only prevented us from attaining but even from expressing. In fact, in objectivistic jargon *'exchangeable* events' would be translated into 'equally probable and independent events . . . with *unknown probability'* (where thanks to the last observation it is ultimately found that *the probability varies* and *there is no independence!*).

Before going on with this reconstruction work, it must be remembered that we have to give an answer to the two remaining questions.

THE 'RELATIVE' NATURE OF PROBABILITY - VII

After eliminating the main cause of confusion deriving from the belief that a given probability must be attributed not to each single event but to collections of events, whether these collections are finite or often even infinite, or worse still, strung in an imaginary infinite sequence (von Mises' "Kollektiv"),[11] let us now go on and examine questions two and three.

The question "in what circumstances" and the (unassailable) answer "all the circumstances known to be relevant at the time", shows that not only does the evaluation of probability have to refer to *an* event (*specifically*) but it also depends on the variable group of circumstances believed to be relevant to its occurrence which are known at the time (and, generally, vary from moment to moment).

In other words, and more precisely, it varies according to the *state of our knowledge* which can be continuously enriched by the flow of fresh information and, among other things, supplied with the results that are gradually learnt or observed in relation to more or less similar situations and cases. In particular, in the case of 'similar cases' it is customary, in the languages criticized, to state that probability *is* the observed frequency (or that the latter "almost certainly approximates to . . .", or "is a good estimation of . . .", etc.). There actually is something valid, if not in the statements themselves, at least in the idea underlying them, even though they are riddled with the contradictions rooted in all objectivistic imitations of probabilistic statements. The same conclusions, translated into subjective form and based on exchangeability, which is the translation into subjective form of the false "independence and equiprobability," provide a natural solution to this vexed question.

(11)   For R. von Mises it would be possible to speak of probability only for an infinite number of events ordered into a series; it would be *defined* there as a limiting case of *frequency* (or percentage of *successes*) in the first n cases when n tends towards infinity. As it is absurd to attempt to make an infinite number of *trials* this precaution is an illusory subtlety aimed at having frequency roughly identified with a finite number of trials.

Quite apart from the peculiarities of any specific case or example, the complete, well-founded argument is based on *Bayes' theorem*, i.e. on the foundations of *Bayesian induction* (or, as it is often called, of *Bayesian statistics*). Here I wish to make it quite clear from the outset why I have an aversion to this term. It is not that I do not subscribe to Bayes' position but because it is the only *correct* one. Anyone who says "BAYESIAN INDUCTION" ought, by the same token, also to call "PYTHAGORIC ARITHMETIC" that arithmetic which follows the traditional PYTHAGORIC TABLE in order to obtain a product, allowing that 6 times 8 is 48, 3 times 9 is 27, etc. (while others, perhaps following a new fashion, might prefer 6 times 8 to be 90 and 3 times 9 to be 77).

In order to limit ourselves for the time being to clarifying how each new item of information modifies our probability evaluations, we shall refer to Euler-Venn's famous "potatoes" (which are very useful, provided they are not overused as certain "fashions" would like).[12] So as to make the explanation more concretely intuitive, let us suppose a table against which an arrow is shot, and that *"success"* consists of hitting a point lying within area E (dotted potato). For convenience, let us suppose the density of probability to be uniform in the sense that the areas have equal probabilities. (This in no way affects the generality of the conclusions, but it is more concrete to say *area*, of course, taking the area of the table as a unit.) $P(E)$ is thus the area of the dotted zone E. However, we have also delimited other zones ("potatoes") marked H_1, H_2, etc. If we are informed that the arrow has struck a point lying in region H_1 (for instance), we should no longer assign the probability $P(E)$ to the "success" but the probability $P(E|H_1) = P(EH_1)/P(H_1)$: ratio between the area of H_1 contained in E and the total area of H_1. If we also were to learn that the arrow has *also* struck, for instance, the region H_3 (and thus the intersection zone H_1H_3) the probability would become $P(E|H_1H_3) = P(EH_1H_3)/P(H_1H_3)$, etc.

This is the sense (or the formal translation of the sense mentioned earlier) in which probability is *relative*, i.e. relative to the *present* state of our information (moment by moment), which constitutes an event H. When we speak of the "probability of E" (with no further specification) and indicate it simply as $P(E)$, it is understood (note well!) that, in evaluating it we are referring to the present state of our *knowledge*. In order to be explicit, we

(12) The familiar term of *potatoes* has the merit of revealing immediately the purely indicative value of this visual representation. By saying (as is customary, unfortunately) that the potatoes are *sets of points* (which are alleged to be . . . *possible elementary cases!*) everything is falsified and made absurd. (Some seem to be embarrassed about saying or not which region the points of dividing or boundary lines belong! Not even at the time of the epic feuds over stolen buckets was there any fighting over the sovereignty of *points* on the boundary *line!*).

should indicate it, for instance, as H_0 and write $P(E|H_0)$ just as, in general, we should indicate as $P(E|H)$ the probability *subordinate* to or conditional on the event (or condition, or "hypothesis") H.

Here too H is generally used to indicate (explicitly) only the possible additional hypothesis (with respect to the H_0 already assumed) by writing $P(E|H)$ instead of $P(E|H \cdot H_0)$. (H_0 is *understood,* not eliminated!)

Referring to the "potatoes" thus means that the whole "table" was already the "potato" which includes only the cases possible at the time it is posited: a moment before it itself would have been a potato in a larger table (i.e. including cases that have already been excluded in the meantime by the arrival of further information).

In betting terms (*the only really and realistically significant ones*) the probability $P(E|H)$ is the *price p* to be paid for a bet that is

		gain		
annulled	$\begin{cases} p \text{ is paid} \\ (\text{resti-} \quad = 0 \\ \text{tution}) \end{cases}$		if H does not occur	E is not important
won	1 is paid $= 1 - p$	$\Big\}$		and E occurs
			if H occurs	
lost	0 is paid $= -p$			and E does not occur

This is the basis and the justification of the inductive argument: as information varies: $H_0, H_1, H_2 \ldots, H_n, \ldots$ increasingly restrictive (the ones that, on commenting the potatoes, we indicates as $H_1, H_1H_2, H_1H_2H_3$, etc., as they were visualized as successive intersections, although this is not essential) probability gradually becomes

$$P(E|H_k) = P(EH_k)/P(H_k) \text{ where } k = 0, 1, 2, \ldots, n, \ldots \text{ etc.}$$

The simplest and most classical case is that of Bayes-Laplace, which corresponds to a uniform mixture of probabilities between 0 and 1. In an approximate version it consists of drawing from an urn containing a large number N of black and white marbles with equal probabilities $1/(N + 1)$ for the $N + 1$ "hypotheses" that the black marbles are $0, 1, 2, 3, \ldots, N - 1, N$, the marbles being returned to the urn after extraction. After *n* extractions, of which *h* are black marbles, the probability of drawing a black marble in any further 'trial' (at that moment, with that information available) is $P = (h + 1)/(n + 2)$ (Laplace's 'rule of succession'): i.e. the one ridiculed by his opponents who applied it to the 'probability that the sun will rise tomorrow'.

It is more interesting to observe that the evaluation in this case is identical with that of another case that *objectively* seems quite different, i.e. that of 'Polya's scheme.' One begins with two marbles in the urn, one black and one white, and successive extractions are carried out, each time returning the marble extracted *plus one of the same color.* It is obvious that after *n* extractions, of which *h* are black marbles, in the urn there are $n + 2$ marbles, $h + 1$ of which are black and the probability of the next extraction is again $(h + 1)/(n + 2)$. Arithmetically speaking, the conclusion is the same; the difference here is that it corresponds to the actual composition at this time, while in the other case, it was not the composition that varied but its evaluation based probabilistically on the observed result.

INDEPENDENCE-EXCHANGEABILITY - VIII

At this stage, before going on to the third and last point (on subjectivity) we could perhaps conveniently make a digression involving a few observations.

Firstly, it must be remarked that the *relative* nature of probability is not usually stressed as being a general fact: sometimes it may be understood, but it usually does not even seem to be noticed. In any case, insufficient attention seems to have been paid to the consequent necessary recommendations (were it only by means of the "DANGER" road sign placed in the page margin *à la* Bourbaki).

One simple and significant example is enough to show how uncertainty can exist in those interested even as a result of the incompleteness of the information available for each of them, while the event (in this case the victory of one of them) is certain, and would be found to be so if only they would exchange the information that each one possesses.

Let us take the example of playing Bingo for a prize among three individuals whereby each extracts a number from the urn (without replacing it), the one drawing the highest number being declared the winner. As long as each one knows only his own number, *m*, he will evaluate as $p_m = K (m - 1) (m - 2)$ $(K = 2/(89.88))$ his own probability of winning and as $(1 - p_m)/2$ that of each of the other two (only for $m = 90$ or 1 or 2 is he certain to have won or lost). Indeed, this is one of those classical probabilities that the objectivists would actually call "objective."

If two exchange information, the one with the lower number knows he has lost. For the other probability rises to $(m - 2)/88$; if all three exchange information, the winner is thus identified.

It thus becomes evident just how unfounded is the notion according to which probability is made to depend (or at least we are induced to believe that it can be made to depend) on an 'uncertainty in itself,' as it were, of the fact considered (or, as is the same thing, of an 'uncertainty of Nature' in connection with it). Nor, in order to eliminate this kind of paradox, would it be enough to allow these evaluations to be modified subjectively (indeed, even if they happened to eliminate the apparent paradox, they would be more absurd than ever, as evaluations based on different information cannot behave in the way that would be correct if they were based on the same information).

Often, as a result of this misunderstanding, especially in rash *axiomatic* approaches, the impression is given that $P(E)$ can be used to fix everything up and that $P(E/H)$ is something ancillary and definable as $P(EH)/P(H)$. Worse still, the exclusive use of $P(E)$ can lead to the objectivistic error in its flattest form, i.e. where the probability $P(E)$ of event E is considered as an objective magnitude that is intimately connected with event E (instead of varying with variations in the understood, but never ignorable and eliminable state of information expressed by the 'hypothesis' H.)

This has a more serious effect, however, on our understanding of the notion of (stochastic) independence. By writing it as $P(E_1 E_2) = P(E_1) \cdot P(E_2)$ (understanding H_0 as is legitimate but dangerous) one could get the idea (or it may seem obvious) that the notion of independence has an absolute meaning rather than one *relative* to this or that information state H, and that the preceding expression should imply also that $P(E_1 E_2 | H) = P(E_1 | H) \cdot P(E_2 | H)$ whatever the value of H. Or at least it might seem 'correct' (as, grammatically speaking, it unfortunately appears to be) that if two events E_1 and E_2 are independent under each of the distinct and exhaustive hypotheses (i.e. forming a partition) $H_1, H_2, \ldots, H_n$, are also 'independent' tout-court.

This is not true, however, as several simple examples are enough to show.

Extractions with return of the marble to the urn having a known composition (e.g. 7 white marbles and 3 black ones) are stochastically independent; if, on the other hand, the composition is not known (if we know, for instance, that there are 7 of one color and 3 of the other, and we assign the same probability, 1/2, to the two hypotheses that there are 7 blacks or 7 whites), it is clear that each piece of information on the result of a new extraction increases the probability assigned to the composition of the color of the marble extracted (and the color extracted with greater frequency will gradually draw ahead). *Independence thus does not subsist*: the correct translation of the nonsensical 'constant but unknown probability' is given by the notion of *exchangeability*. In these cases, later extractions are (not independent but) *exchangeable* in the sense that the probability does not vary for permutations e.g. every series of 9 extractions (e.g. of 5 white and 4 black) has the same probability (invariance with respect to the permutations). For instance:

$$P\ (N\text{-}B\text{-}N\text{-}B\text{-}B\text{-}B\text{-}N\text{-}N\text{-}B) = P\ (B\text{-}B\text{-}B\text{-}N\text{-}B\text{-}N\text{-}N\text{-}B\text{-}N) = \text{etc.}$$

(for each of the $(4)_9$ = 126 permutations), (where N stands for black and B stands for white).

It is this very *non-independence* that allows us to make the probability evaluation of future cases that is based on the frequency of observed cases under conditions usually, although incorrectly, called of 'equiprobability and independence' (which makes the conclusion patently contradictory). The condition which translates the above nonsense into a meaningful form is "exchangeability": i.e. a property that is maintained for *mixtures* (i.e. for positive coefficient linear combinations) and is valid for equally probable independent events and thus for their mixtures. Indeed, exchangeability is found to be valid for all of these kinds of mixture and solely for them. It is thus a correct way of expressing what was absurdly called 'equally probable independent trials.'

We shall merely state the result without going into technicalities lying beyond our present scope: every mixture of 'repeated trials' having 'constant but unknown probability' (as would be said incorrectly but traditionally and thus 'understandably') leads to an exchangeable process, and *vice versa*; i.e. an exchangeable process can always be interpreted as such a *mixture*. This conceptual clarification, illustrated here in the most banal of examples that, however, can be and is extended to many complex cases, is what I am interested in as it helps to dissipate the concepts (or at least the terminology) of a superstitious flavor, of alleged metaphysics, of contradictory expression. Even though, for those who are more emphatically 'mathematicians' (for whom mathematics is an *end*, not a *tool*) what is more important is the analytical result they have christened *"de Finetti's representation theorem."*

PROBABILITY: ITS 'SUBJECTIVE' NATURE - IX

As a last step, as the last link in the chain of observations that had gradually excluded mystifying interpretations of the idea of probability, we should add other remarks to demonstrate the basically and necessarily *subjective* nature of probability. But this is now tantamount to flogging a dead horse as a probability concerning *events* in the sense of 'single events' and depending on a state of information of the subject evaluating it is already basically 'personalized.' The only additional observation to be made is that a 'personal' component will certainly be present to differentiate a number of evaluations made—also on the basis of a supposedly identical state of knowledge—by different individuals giving greater attention or assigning greater weight to certain circumstances than to others. This will be true quite apart from distorting factors like 'wishful thinking' or 'painful thinking,' emotional reactions to certain details, etc.

However, I am too much of a subjectivist (and a pragmatist) to believe there is anything to be gained by starting up a metaphysical-verbalistic diatribe. Each individual making a *coherent* evaluation of probability (in the sense I shall define later) and desiring it to be 'objectively exact,' does not hurt anyone: everyone will agree that this is his subjective evaluation and his 'objectivistic' statement will be a harmless boast in the eyes of the subjectivist, while it will be judged as true or false by the objectivists who agree with it or who, on the other hand, had a different one. This is a general fact, which is obvious but insignficant: *"Each in his own way."*

However, it must be stated explicitly *how* these subjective probabilities are *defined,* i.e. in order to give an *operative* (and not an empty verbalistic) definition, it is necessary to indicate a procedure, albeit idealized but not distorted, an (effective or conceptual) experiment for its *measurement.*

Nothing has to be invented; it is sufficient to reverse a statement made by objectivists when they *define* a *fair* bet as one where, in order to receive a sum S (either positive or negative) if event E with probability p is found to be true, the payment pS is required. In this version the *probability* is thought of as being defined *first,* from which the *fair* bets are deduced.

Clearly, however, this means putting the 'cart before the horse': we must imagine an individual of being in the condition, or rather of *having,* to bet on the occurrence of a certain event E, and see which bets on E he would judge to be *fair* (i.e. acceptable, in his opinion, in both senses indifferently). If, for any S, positive or negative, he considers as fair an exchange between a certain sum pS and the right to a sum S contingent on the occurrence of event E, this coefficient p is known as the probability of E (for this individual) and is denoted $p = P(E)$. More roughly, p is the *price* that this individual assigns to one dollar dependent on the occurrence of E; and by thinking of prices, all becomes clear (and even banal).

The important thing to stress is that this is in keeping with the basic requirement of a valid definition of a magnitude having meaning (from the methodological, pragmatic and rigorous standpoints) instead of having remained at the level of verbal diarrhoea: it must not be based on vain or over-elaborate phrases, but it must be *operative,* i.e. based on the indications given by experiments, albeit conceptual ones, that must be carried out in order to measure it.

This relates back to the brief mention made of Calderoni and Vailati, as well as to the positions regarding Mach, Einstein, Bridgman, etc. And everything becomes clear: all the customary rules of the theory of probability are derived as simple corollaries of the additive necessity of *prices*. For 'lottery tickets' (in the broad sense: wages on any facts and for any amount) the price is established as for any goods or basket of goods: $\Sigma_h \, p_h \, S_h$ is the price in order to receive $\Sigma'_h \, S_h$ (indicating with Σ' the sum restricted to *actual* events). This sum could more appropriately be indicated by $\Sigma_h \, E_h \, S_h$ (which indicates the same thing more significantly as it has been agreed that $E_h = 1$ or $E_h = 0$ according to whether E_h is *true* or *false*. (Incidentally, this is an example of the use of indicating true and false by means of 1 and 0 and the usefulness of such a convention).

In this way light is shed on the true, simple meaning of all the 'laws' or 'rules' of the theory of probability: they boil down to the *consistency* (in the sense of *additivity*) necessary for the prices of any kind of goods or objects (including 'lottery' tickets).

In English, a combination of bets devised in such a way that, profiting by an inconsistency in the odds given by the bookmaker, someone is certain to win *whatever happens* is called 'Dutch Book' (I don't know why). However, if one wants to, this term could be used to express the condition of consistency that is the sole basis on which the whole theory of probability rests: suffice it to say that it consists in *allowing no chance of a Dutch Book occurring*.

HOW TO 'EXPLORE' PROBABILITY - X

We can now answer a question that should actually be considered as *prejudicial*: how can one *know* the probability $P_A \, (E)$ that a given individual A (in a given instant, at the present state of his information, understood) assigns to event E?

In general, he would not be able to tell us (even if he wanted to, unless he has had enough practice). And even if he were to tell us we should not be certain of his sincerity (he might want to trick us, or just answer anything, as one does to pests bothering one with questions that one finds silly or indiscreet, etc.). And not even if we had some information about his sporadic betting would it be much use to us: he might bet capriciously, to 'try his luck,' without making any revealing reflection.

On the other hand, suitably devised 'mini-bets' can be revealing reflections, if the subject is interested in collaborating. He is asked to indicate the odds p on the strength of which he would be prepared to accept a 'mini-bet' devised in such a way as to render excellent, i.e. as advantageous as possible, in his opinion, the sincere answer in conformity with its effective evaluation $p = P(E)$.

To be precise, on the basis of this value p that he himself has indicated he will be *penalized* by $(1 - p)^2 \, S$ if E occurs and by $p^2 \, S$ if 'non-E' occurs.

In practice, in order not to take advantage of his willingness to collaborate by inflicting penalizations (i.e. pure losses) on him, the sum S could be offered to him so that he would lose only a part of it (i.e. in both cases he would win $[1 - (1 - p)^2] \, S$ or $[1 - p^2] \, S$; he would lose 'everything' only in the limiting cases: (proffered) probability 0 and the event occurring, or (proffered) probability 1 and the event not occurring). The entire theory of probability derives from the latter premise (which is not an *axiom,* or anything artificial or crazy or sublime but a condition of patent common sense). In my view, this is the most important and significant circumstance and I believe it must be explained as explicitly as possible. To clarify something to the extent of showing that it is obvious may seem to be disrespectful for anyone fond of pomp, rhetoric and the mysterious, but I consider it to be the definitive attainment of a conclusion in its optimal form. In this way, and only in this way, can one reach a *true* definition that is not deformed by chatter or pseudometaphysical or abstractionistic tendencies, but is *operative, behaviorist* and *pragmatic,* based not on pretentious words but on conceptually (and actually) experimental choices.

The one described here is only one (the simplest and most significant) of the personalization rules that are 'suitable' in the sense defined above (proper scoring rules). It is useful, however, to outline the demonstration and to add further considerations and information.

For the individual making the evaluation, it is immediately apparent that the best thing is to express his opinion exactly and sincerely because this is the only way to reduce the forecast of the penalization to a *minimum* (according to the estimate of the person sharing this evaluation, and therefore according to himself).

If, in fact, one claims to have chosen the probability p but in actual fact assigns a different value, $\bar{p}$, to the probability, the predicted penalization is, in his own opinion:

$$\bar{p}(1 - p)^2 + (1 - \bar{p}) p^2 = (p - \bar{p})^2 + p(1 - \bar{p}) = (\bar{p} - \bar{p})^2 + \text{const.}$$

(referring to p), and becomes a minimum only if it is said that $p = \bar{p}$ (i.e. if the sincere answer is given).

This rule (Brier's) is applied in the USA to indicate the probability of rain in the meteorologists' evaluation in radio and TV weather bulletins as well as (also quite independently by us in the University of Rome) for probabilistic predictions on the results of football matches (ever since 1959). The first to suggest these methods (and who remained the unknown forerunners for some time) seem to have been the American McCarty and the Japanese Masanao Toda, during the 'fifties. A feature of special merit is the fact that the quadratic penalizations, on being summed, always give (the square of) a distance to be minimized in a space with a number of dimensions equal to the 'degrees of freedom' (i.e. to the linearly independent evaluations: here, two dimensions for each forecast of a match result). In particular, this is related also to the mechanical significance of the center of gravity as a point in which the moment of inertia is a minimum; but I shall not go into this deeply.

What I am more concerned with is to stress the practical importance of using these methods as they provide a way of giving precise expression to judgments that, in words, are extremely ambiguous and thus not serious and relatively useless. For instance, in oil research (as is amply demonstrated in a book by C. J. Grayson on the subject) the answer of an expert expressed in the terms of an evaluation of the probability of finding oil in a given place (and in the prediction of the quantities in which it exists) is much more useful and serious than vague, qualitative and elusive words, often hovering between saying and gainsaying, and sometimes as hard to interpret as the oracles of a sibyl. Answers couched in the terms of probability and prediction can be used as the basis for estimate calculations (even taking account of the degree of approximation and indeterminancy they cannot help containing) and increasing the confidence with which the oil-men and the experts collaborate. Quite apart from the utility of the practical applications, the use of these methods is precious in itself as a factor in the psychological refinement needed by everyone, as actual or potential makers or users of this kind of evaluation, in order to create the habit of *seeing* the connection between degree of uncertainty and its numerical expression as a probability.

This kind of training would, in my opinion, be very useful in all schools, both in the probabilistic field itself and in order to refine one's capacity for the numerical appreciation of all other kinds of magnitudes, e.g. size of a crowd, distance, area, time duration, weight, temperature, speed, etc.

Unfortunately, this educational factor seems to be rather neglected. In this connection, I can only think of two examples (I hope this is due to my poor memory): a few estimates of lengths and distances (with an empirical but instructive discussion of the distribution of errors) carried out by eye for the same length by different pupils in Mario Lodi's school (cf. *Il paese sbagliato*, Einaudi, Torino, 1970; quoted in Periodico de Matematiche [PdM], 1973, n. 4), and the true training of the Indian boy Kim (in the novel of the.

same name by Rudyard Kipling) so that he could more effectively act as an informer during the British *raj* in India.

THE AMENDMENT OF STANDARD CONCEPTS - XI

The main outline of my exposition could now be said to be concluded only that it would be a little too dry and unassimilable if we left out all reference to connections and comparisons with habitual ideas and expressions.

These observations are of three kinds and I should like to say at the outset that "I shall be brief," although, as in the case of Ophelia's father in Hamlet and of many of my colleagues in countless asphyxiating academic council meetings makes me fear that this phrase will sound alarming to the reader's ears.

First of all, I should like to warn the reader against D'Alembert's advice that would be disastrous if applied in the field of probability (as also, probably, in that of Analysis to which he was referring) i.e. "*Allez de l'avant: la foi vois viendra.*" If one goes ahead without freeing himself, by clarifying them, of the first natural misunderstandings or reservations, it will certainly be true, unfortunately that faith will come to him but not in the correct sense of "I believe because *it is clear*" but in the aberrant one of "*credo quia absurdum.*"

This is tantamount to an acceptation of the notion of Probability as a *deus ex machina* deriving from abstract arguments that carefully avoid explaining that the meaning of probability is the one that all of us (human beings and other animals) use to evaluate dangers and risks and prospects of varying degrees of rosiness. Indeed, probability is our *guide* when we *think* and *act* in conditions of *uncertainty*, and uncertainty is *ubiquitous*.

The theory of probability is the logic (which is more or less instinctive and perfected as an instinct and rational faculty, either unconscious or more or less scientifically organized and ingrained) with which we make our choices for the purpose of optimizing our chances.

Probability calculus allows these unconscious and instinctive arguments to be translated into a scheme of careful evaluation of the *pros* and *cons* which could hardly (nor I think should it) replace the spontaneity of instinctive decisions with cold profit-and-loss accounting. It should rather be of use in perfecting and controlling this spontaneous gift and in supporting it by means of significant orientative indications. To what extent can traditional definitions of probability be useful for this purpose? According to the first one it is the ratio m/n between the number of successful and possible cases '*if they are equally probable*'. For us, as probability has been defined as the price of a bet and as it must thus be additive, this is *true and obvious*, not however as a definition but as a *corollary*. If *p* is the probability, that someone admits as being *equal*, of each of the *n* cases (incompatible and exhaustive), the certain event (their sum) has a probability of $np = 1$: so $p = 1/n$ and for a sum of *m* such events the probability is m/n.

Someone would perhaps insist that this probability *is* therefore objective, but this is not so: the very judgment of equal probability is itself subjective. It may further be insisted that if *everybody* (or all *reasonable* people) share this evaluation it would mean that it was objectively correct. But, in addition to the vicious circle set up by deriving the reasonableness of an assertion by postulating the reasonableness of those accepting it, a sum of so many subjective opinions cannot lead to an objective conclusion: it would be like thinking that by increasing the number of stones in a heap it would finally become an animal.

If our contradictor falls back on the statement that, in any case, an opinion shared by many is *reasonable*, or *natural*, I declare myself to accept this concept on the whole; although it merely refers to a probable acceptation of what many others also accept subjectively on the strength of similar motives and nevertheless remains a *subjective* opinion (albeit 'multisubjective' or 'intersubjective').

I too can use and recommend using terms like 'reasonable' and 'natural' in this sense, not without stressing however that the sense is by no means objective but *subjective to the second power*: (subjective^2), as it is a subjective opinion about other people's subjective opinions.

The intrinsic contradictions in claims to 'objective' definition or explanation can always be shifted elsewhere but never eliminated: it can be said of them (to use the apt image suggested by Bernard O. Koopman[13]) that *"unlike Napoleon's Guard they always retreat but never die."*

This is true (even more so) of the analogous situation concerning probability evaluations based on observed frequencies (frequentist or 'statistical' conception of probability). To relate probability back to frequency has no meaning if the meaning and conditions are not specified. (This we shall do shortly when the notion of *exchangeability* announced right at the beginning is presented). Let us just make one advance observation (so as not to interrupt the treatment later): all that has been said previously (para VI) to denounce the equivocal implications of the expression 'trials of one and the

---

(13) Suffice it to quote as an example an excerpt from *The Bases of Probability*, in Kyburg and Smokler, *Studies in subjective probability*, (pp. 161-172). *"For the purpose of concluding that* A *is irrelevant to the occurrence of* E *on any given occasion, on the basis of the results of experiments carried out on a different occasion, it must be assumed that certain other inevitable differences between the two occasions are themselves irrelevant to the situation being considered. The difficulty, quite unlike Napoleon's Guard, always retreats but never dies"* (p. 172).

same event' already contains sufficient grounds to extend to the new attempt, and more drastically, the rejection of any claims of being able to transform, probability into something 'objective' in this other way as well.

An attempt to allow different notions of probability to coexist was made by Cantelli who recognized three different 'schemata': the schema of the *urn* (i.e. that of the 'equally probable cases'), the *insurance* schema (based on the frequencies given by the statistics[14]), and the *horse* schema (referring to betting at race meetings).

I should say that these kinds of distinctions (these or others) can be made, but concern comparatively external aspects of the type of information and do not detract from the unitary nature of probability in all fields. The difference, to express it by means of an analogy, is the same as that between lengths measured with a tape measure or geodetic instruments, or radar signals, or in any other way: there are many methods but the magnitude is always a distance (perhaps measured with varying degrees of accuracy but always and only a distance).

THE TOWER OF BABEL - XII

The confusion (or semiconfusion) between probability and frequency and others mentioned, to be mentioned or that we shall refrain from mentioning, would seem to indicate that the Tower of Babel really existed and that this is its wretched heritage. Here we shall resume and conclude the criticism of the objective reasoning and doubts about frequency (which have already been solved, para VIII), to indicate to what extent obstinate misunderstandings are sometimes the inevitable result of unhappy and misleading expressions. The expression is that of the events known as 'trials of one and the same event', which are supposedly 'independent' and 'equally probable', but *'with unknown probability'*. Benevolently reading between the lines it can be understood that this is the definition of exchangeability. Indeed, this is what it is alluding to but in an irremediably contradictory way (since the probabilities are depending upon the observed outcomes). The definition of exchangeability is needed to dissolve ambiguities and contradictions together

(14)   In connection with this example, in order to avoid *superstitious* interpretations, it must be observed that death rate tables are used only as a standard basis, as a first reference, for individuals of a given age. But the insurers ascertain all the specific circumstances that, for each individual, may influence, especially negatively, the probability of death, for the purpose of making a *personalized* evaluation. If, however, in practice, superstandard premiums are avoided so long as the profit margin allows it, it is because it is preferable to gain a client rather than risk losing him.

with the terms generating them. A basic confusion derives from the assertion that probability ONE and probability ZERO are the same, respectively, as CERTAINTY and IMPOSSIBILITY *in the logical sense*: it would be like stating that a set of zero measure (either in the sense of Jordan-Peano or of Borel-Lebasgue) can only be an empty set!

If anything one could speak of ALMOST-CERTAINTY and ALMOST-IMPOSSIBILITY (with 'almost' being understood in a similar way to its use in Analysis, e.g. in saying 'almost-everywhere').

It is more useful, however, to think of the simpler and more decisive example of a uniform probability distribution over a certain range (e.g. 0, 1). Obviously each point has zero probability, although the set of *possible* points is necessarily *infinite,* since it must be dense everywhere (perhaps only enumerable, as for instance the set of rationals, or with the power of the *continuum* like the whole line or the subset of the irrationals).

The example of the rationals, however, leads to a much more important conclusion, i.e. that the *complete additivity* (or σ-additivity) is a property that probability does not necessarily possess. (Moreover, also in the theory of measurement this property may be shown to be valid only by limiting oneself, e.g., to the sets measurable according to Lebesgue). And it is the result of the custom of using such a doctored field that (unjustifiably) makes one think that what happens outside the artificial enclosure is absurd.

But the same confusion between 'certain' and 'almost-certain' is found in many other situations as an inexhaustible source of aberrations. Too often one erroneously says 'certain' and 'impossible,' not only when they have probabilities of '1' and '0' but also when these probabilities merely approach '1' or '0'. It is this inconceivable and unpardonable bad habit that leads to such absurdities as that of 'defining' probability as 'frequency,' 'irregularity' (randomness) as a *necessary* characteristic of the successions obtained by chance (even to the extent of sometimes taking it as a 'definition' (sic!!!), and to many other similar ones. A few sometimes even seem to think that it is necessary to believe, and make others believe, that a magic exists whereby this is produced (or the opposite) concealing the extremely commonplace fact that explains everything by deflating the balloons: the Chance that everyone speaks of certainly has no predilection for series with frequencies equal to or approaching the percentage of black or white marbles (in the case of extractions; the same is true for any type of 'repeated trials' in the same hypotheses). Chance (with a capital 'C') does not choose certain series because they are the ones it prefers but is dependent on 'chance' (with a small c) which chooses blindly, with no predilections, and in general the series most frequently occurring are of the type that is most numerous.

And yet, among the other things, many find it strange that there should be long sequences (many consecutive extractions of a black marble, or, to mention the most pitiful case, of lottery extractions during which a 'retarded' number is not drawn!) But one must not be too hard on the poor devils who believe in such a superstition even to the point of ruining themselves, as we all seem unconsciously to be more or less affected by 'reluctance to the unusual'. According to Komlòs and Tusnàdy ("On sequences of pure heads," *Ann. Prob.*, 1975), this derives from the observation that, when someone is requested to write a long series of H and T to exemplify (or 'simulate mentally') a heads and tails process, everyone is generally found to fall systematically into the error of never indicating, or much more rarely than is indicated by the theory and found in practice, 'long' series of H's or T's alone. (According to Rényi, [1970], the length of the longest 'pure' succession out of n attempts (with diverging n) ought to be of the order of magnitude of $loglog\ n$.) Anyone pretending to impersonate Chance seems to be concerned with imitating it *too well* and fails for this very reason: he does not imitate Chance because Chance has no memory and succeeds in creating random series because it does not set out to, or even know that it is doing so.

DETERMINISM AND INDETERMINISM - XIII

The doubts concerning the alternative: 'Determinism or Indeterminism' are of a deeper nature. I think that the alternative is undecidable and (I should like to say) illusory. These are metaphysical diatribes over 'things in themselves'; science is concerned with what 'appears to us,' and it is not strange that, in order to study these phenomena it may in some cases seem more useful to imagine them from this or that standpoint, by means of deterministic theories (and allowing for 'accidental errors' or 'inaccuracy of measurement') or indeterministic ones (and the explanation virtually finds an advance explanation in Castelnuovo's treatment). It is not possible, in fact, to discern any process of propagation deterministically idealized as the resultant of elastic collisions between ball bearings, or translated into a process of 'diffusion of a probability,' as it were, of the Wiener-Lévy type. Incidentally, I should be very careful not to say that it is a 'probability density' that is being propagated or bouncing or being absorbed by reflecting or absorbing barriers, etc.; if anything, I should make a point of saying it 'in inverted commas,' or pointing out that 'everything occurs *as though* ... (etc.,),' in the spirit of Veihinger's *'als ob'*.

Even worse, because it is completely superfluous and futile, is the contortion usually used even now to apparently increase the 'objectivity' of the statement that "we can expect with a very high degree of probability that

... etc.," by saying that in a very large collection (called *'ensemble'* in thermodynamics, according to Gibbs) "in the vast majority of them" it happens that etc. (in the hope, it appears, that this emptily hypothetical phrase will seem to be more 'objective' than the simple statement that "we can expect with a very high degree of probability that ... etc.").[15]

An example may also be taken from mathematics. The value of π is certainly well-defined (i.e. all the infinite decimal figures are: 3.141592 ... whether already calculated, or not yet, or never by man). And yet, as there are no reasonable hypotheses to the contrary, i.e. abnormal with respect to 'random choice' (prob. 1/10 for every figure, quite apart from the preceding ones) Borel has concluded (and I think correctly) that for us mortals it is reasonable to assign a probability $= 1$ (even though it is not a *certainty* but an 'almost-certainty') to the hypothesis that π is a 'normal' number for decimal notation (and indeed 'absolutely normal', i.e. normal for all number systems and not only the one based on 10) as the numbers that are not 'absolutely normal' are 'very few' in number (they belong to a zero-measure set).

I shall not add any further examples or observations, but merely point out the care needed to separate what *has meaning* from what *has no meaning* (according to the recommendation of Calderoni and Vailati) or, to say it with a maxim taken from the Gospel, to carefully separate "the wheat from the chaff." In our case, to separate the wheat from the chaff means reflecting on the different way in which phrases, statements and conclusions concerning the same problem are formulated and correlated with reality according to whether a subjectivistic or objectivistic conception of probability is adopted.

PROBABILITY, MATHEMATICS, TEACHING - XIV

In order to conclude constructively, there remains one more question to be dealt with on which I should like everyone to reflect seriously.

I find there is a dangerous, detrimental and deplorable tendency in both the scientific and the teaching fields toward unilateral, sectorial, and

(15) It is odd how some people, by saying that they *understand probability in an objective sense* (without explaining the non-existent sense of this statement) manage to have the assertion accepted as something meaningful by other people. Unless it is something else, i.e. *objectivity,* sincerity in evaluating dispassionately all the circumstances (always, of course, subjectively), without deceiving oneself (out of pessimism or optimism) or others (to cheat them or to favor or oppose their preferences, etc.).

reductive, often even artificial conceptions concerning what Probability *is* and what it *can* and *must be used for.*

The considerations I wish to make here refer in varying degrees to both Probability research and teaching at all levels, ranging from the pre-school, primary, secondary and tertiary educational levels up to the university and that of specialists and publications in the field, the latter too often under the pathological spur of 'publish or perish'.

On the whole it can be said to be a *generalized* bad habit, i.e. common to the whole of mathematics teaching. However, in our field, it is joined by other ones peculiar to probability theory and the fields in which it is applied (statistics, operations research, etc.).

The generalized bad habit afflicting mathematics teaching is that of putting 'the cart before the horse' (as was stated and aptly illustrated by Gemma Harasim, the mother of Lucio Lombardo-Radice, in an article written in 1915, and partially reprinted in PdM, 1975, n. 1-2). Dispensing with metaphors, the error is to 'TEACH' mathematical formalism without first having made the need for it felt or having explained its meaning: for this reason it is essential to explain 'what it is used for' (of course, not only in the commonplace meaning of 'doing sums', and this can be vividly revealed to both children and uneducated people).

As for the *specific* bad habits, I think five can be listed: I shall then deal with how to avoid them (thus allowing dissenters the possibility of saying what 'my bad habit' is).

First. If we start from common cases and things where the evaluation of a probability is (or at least seems to be) obvious, mechanical and perhaps the independence is too, etc., the risk, consists of thinking that everything, or almost everything, can be reduced to combinatory calculus. This would be like introducing (or perhaps defining) *area* only for rectangles and triangles, allowing the distorted idea to be formed that the notion has no meaning for other figures. Combinatory calculus is certainly one of the tools often used to solve many problems in probability calculus but it is not in itself Probability Calculus.

Second. If a certain type of 'logical' formalization is followed (where 'logic' is no more than a formalistic symbolism) one remains more or less at the same level, which is however further obscured and hidden by arbitrary pseudological conventions and by the intention of considering 'probabilities' not of *facts* but of 'propositions', and thus not of a concrete but of a metaphysically-oriented flavor. The only possible result would be to convince people that all conventions are good when, and only when, they are passed off as 'axioms'.

Third. I have already made implicit mention of certain deficiencies (para VII) of the 'axiomatic' approaches, where, with a certain responsible attempt at reasonableness, things interpretable with reference to the albeit unnominated 'concrete' are postulated as properties of the function $P(E)$. These deficiencies are easy to eliminate but symptomatic of lack of awareness. The main problem is a different one: one remains at an abstract level of mere *measurement theory* until a real meaning is associated with the symbols and operations and one does not show up which interpretation and value are to be given to the conclusions: − subjective? − or supposedly objective? and if so, pray, − in what sense?

Fourth. Other approaches lead immediately on to a consideration of particularly interesting problems with aleatory (or 'stochastic') problems of the 'Markov's chains' type, to give but the most elementary example, and end up by virtually studying a process of propagation or something of the sort under various hypotheses, stating that what is propagated is a *probability* but not what is meant by Probability and so not what in practice is the result. Or else this appears confusedly in an 'intuitive' sense, which is muddled and mysterious at the time, as it appears as a *deus ex machina* instead of as a clarifying premise set out right from the beginning. And in this case, one can give a fine course of exercises of Analysis (finite difference equations, or differential, partial derivative, integral or integro-differential equations: or Fourier transforms, etc.), almost as an end in itself.

Fifth. Teaching and introductions based on Statistics (or specific branches of it, e.g. Operations Research, Information Theory, demographic and actuarial applications, etc.) are extremely numerous. A certain degree of confusion is often made between probability and frequency but the most specific fault to my mind is the overfondness of numerous statisticians (I do not wish to over-generalize and include in this criticism those who are untainted by this flaw) for rough, empirical methods like those (as I mentioned earlier) Irving Good called 'adhockeries,' while others speak more rudely of 'statistical cook-books.'

It is my impression that those who are untainted are those who have to deal with concrete problems, as in Operations Research, Reliability Theory, etc. and I hope that for once (in spite of the saying of the rotten apples) that the good examples will be more *contagious* than the bad ones. In all five of the cases listed, although for different reasons and from different standpoints, the Theory of Probability shows itself, sometimes deliberately and in any case always *de facto,* to be a doctrine in which everything is known about Probability and its formal properties, but, as in the statement we quoted by Birkhoff, no one knows what it is, nor is concerned with knowing it, or perhaps they want to keep it a secret.

It would seem that one must think, as Hans Freudenthal[16] jokingly wrote, that everyone is making the greatest possible effort to avoid showing Probability "as God created it" by providing it with a fig-leaf, or with so many that it is completely covered, perhaps for fear of being accused of indecency by an over-zealous magistrate.

In conclusion I should like once more to stress that what I am mostly concerned about, the danger to which, on this occasion more than ever before, I have wished to draw your attention, consists of this spate of deviations that are to be found at all levels, both in teaching and in research.

I whole heartedly thank you for your attention and I wish to apologize for taking advantage of it under the urge to say, albeit concisely, all or almost all of what I consider to be essential.

B. de F.

Translation: I. McGilvray - Roma

(16)   This image was taken from an article by Hans Freudenthal (*The Crux of Course Design in Probability,* in Educational Studies in Probability, Reidel ed. Amsterdam, 1974, n. 5: pp. 261-277), which I summarized and commented (also adding other considerations) in *Il buon senso e le foglie di fico: Hans Freudenthal sull'insegnamento della Probabilità,* Boll. Un. Matem. Ital., 1975, Suppl. 3 (in honor of Enrico Bompiani, to whom I had submitted the manuscript to make sure he did not disapprove the jocular style although I was certain he would not as I well knew and admired his intelligent broadmindedness).

RICHARD C. JEFFREY

Probable Knowledge
(1968)

Reprinted from *The Problem of Inductive Logic,* Imre Lakatos (Ed.), North-Holland Publishing Company, Amsterdam, 1968, by permission of the publishers and the author.

The central problem of epistemology is often taken to be that of explaining how we can know what we do, but the content of this problem changes from age to age with the scope of what we take ourselves to know; and philosophers who are impressed with this flux sometimes set themselves the problem of explaining how we can get along, knowing as little as we do. For knowledge is sure, and there seems to be little we can be sure of outside logic and mathematics and truths related immediately to experience. It is as if there were some propositions—that this paper is white, that two and two are four—on which we have a firm grip, while the rest, including most of the theses of science, are slippery or insubstantial or somehow inaccessible to us. Outside the realm of what we are sure of lies the puzzling region of probable knowledge—puzzling in part because the sense of the noun seems to be cancelled by that of the adjective.

The obvious move is to deny that the notion of knowledge has the importance generally attributed to it, and to try to make the concept of belief do the work that philosophers have generally assigned the grander concept. I shall argue that this is the right move.

1. A PRAGMATIC ANALYSIS OF BELIEF

To begin, we must get clear about the relevant sense of 'belief.' Here I follow Ramsey: 'the kind of measurement of belief with which probability is concerned is . . . a measurement of belief *qua* basis of action.'[1]

Ramsey's basic idea was that the desirablity of a gamble G is a weighted average of the desirabilities of winning and of losing in which the weights are the probabilities of winning and of losing. If the proposition gambled upon is A, if the prize for winning is the truth of a proposition W, and if the penalty for losing is the truth of a proposition L, we then have

$$prob\ A = \frac{des\ G - des\ L}{des\ W - des\ L}. \qquad (1)$$

Thus, if the desirabilities of losing and of winning happen to be 0 and 1, we have *prob A = des G*, as illustrated in Figure 1, for the case in which the probability of winning is thought to be $\frac{3}{4}$.

---

(1) Frank P. Ramsey, 'Truth and probability,' in *The Foundations of Mathematics and Other Logical Essays*, R. B. Braithwaite, ed., London and New York, 1931, p. 171.

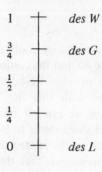

Figure 1.

On this basis, Ramsey[2] is able to give rules for deriving the gambler's subjective probability *and* desirability functions from his preference ranking of gambles, provided the preference ranking satisfies certain conditions of consistency. The probability function obtained in this way is a probability measure in the technical sense that, given any finite set of pairwise incompatible propositions which together exhaust all possibilities, their probabilities are non-negative real numbers that add up to 1. And in an obvious sense, probability so construed is a measure of the subject's willingness to act on his beliefs in propositions: it is a measure of degree of belief.

I propose to use what I take to be an improvement of Ramsey's scheme, in which the work that Ramsey does with the operation of forming gambles is done with the usual truth-functional operations on propositions.[3] The basic move is to restrict attention to certain 'natural' gambles, in which the prize for winning is the truth of the proposition gambled upon, and the penalty for losing is the falsity of that proposition. In general, the situation in which the gambler takes himself to be gambling on A with prize W and loss L is one in which he believes the proposition

$$G = AW \vee \bar{A}L.$$

If G is a natural gamble we have $W = A$ and $L = \bar{A}$, so that G is the necessary proposition, $T = A \vee \bar{A}$:

$$G = AA \vee \bar{A}\bar{A} = T.$$

(2)   "Truth and probability," F. P. Ramsey, *op. cit.*
(3)   See Richard C. Jeffrey, *The Logic of Decision*, McGraw-Hill, 1965, the mathematical basis for which can be found in Ethan Bolker, *Functions Resembling Quotients of Measures*, Ph.D. Dissertation, Harvard University, 1965, and *Trans. Am. Math. Soc.*, 124, 1966, pp. 293-312.

Now if A is a proposition which the subject thinks good (or bad) in the sense that he places it above T (or below T) in his preference ranking, we have

$$prob\ A = \frac{des\ T - des\ \overline{A}}{des\ A - des\ \overline{A}}. \tag{2}$$

corresponding to Ramsey's formula (1).

Here the basic idea is that if $A_1, A_2, \ldots, A_n$ are an exhaustive set of incompatible ways in which the proposition A can come true, the desirability of A must be a weighted average of the desirabilities of the ways in which it can come true:

$$des\ A = w_1\ des\ A_1 + w_2\ des\ A_2 + \cdots + w_n\ des\ A_n, \tag{3}$$

where the weights are the conditional probabilities,

$$w_i = prob\ A_i/prob\ A. \tag{4}$$

Let us call a function *des* which attributes real numbers to propositions a *Bayesian desirability function* if there is a probability measure *prob* relative to which (3) holds for all suitable $A, A_1, A_2, \ldots, A_n$. And let us call a preference ranking of propositions *coherent* if there is a Bayesian desirability function which ranks those propositions in order of magnitude exactly as they are ranked in order of preference. One can show[4] that if certain weak conditions are met by a coherent preference ranking, the underlying desirability function is determined up to a fractional linear transformation, i.e., if *des* and *DES* both rank propositions in order of magnitude exactly as they are ranked in order of preference, there must be real numbers a, b, c, d such that for any proposition A in the ranking we have

$$DES\ A = \frac{a\ des\ A + b}{c\ des\ A + d}. \tag{5}$$

The probability measure *prob* is then determined by (2) up to a certain quantization. In particular, if *des* is Bayesian relative to *prob*, then *DES* will be Bayesian relative to *PROB*, where

$$PROB\ A = prob\ A\ (c\ des\ A + d). \tag{6}$$

(4)   Jeffrey, *op. cit.*, chs. 6, 8.

Under further plausible conditions, (5) and (6) are given either exactly (as in Ramsey's theory) or approximately by

$$DES\ A\ =\ a\ des\ A + b, \tag{7}$$

$$PROB\ A\ =\ prob\ A. \tag{8}$$

I take the principal advantage of the present theory over Ramsey's to be that here we work with the subject's actual beliefs, whereas Ramsey needs to know what the subject's preference ranking of relevant propositions would be if his views of what the world is were to be changed by virtue of his having come to believe that various arbitrary and sometimes bizarre causal relationships had been established via gambles.[5]

To see more directly how preferences may reflect beliefs in the present system, observe that by (2) we must have $prob\ A > prob\ B$ if the relevant portion of the preference ranking is

$$
\begin{array}{c}
A, \quad B \\
T \\
\overline{B} \\
\overline{A}
\end{array}
$$

In particular, suppose that A and B are the propositions that the subject will get job 1 and that he will get job 2, respectively. Pay, working conditions, etc., are the same, so that he ranks A and B together. Now if he thinks himself more likely to get job 1 than job 2, he will prefer a guarante of $(\overline{B})$ not getting job 2 to a guarantee of $(\overline{A})$ not getting job 1; for he thinks that an assurance of not getting job 2 leaves him more likely to get one or the other of the equally liked jobs than would an assurance of not getting job 1.

2. PROBABILISTIC ACTS AND OBSERVATIONS

We might call a proposition *observational* for a certain person at a certain time if at that time he can make an observation of which the *direct* effect will be that his degree of belief in the proposition will change to 0 or to 1. Similarly, we might call a proposition *actual* for a certain person at a certain time if at that time he can perform an act of which the *direct* effect will be that his degree of belief in the proposition will change to 0 or to 1. Under ordinary circumstances, the proposition that the sun is shining is observa-

(5)   Jeffrey, *op. cit.*, pp. 145-150.

tional and the proposition that the agent blows his nose is actual. Performance of an act may give the agent what Anscombe calls[6] 'knowledge without observation' of the truth of an appropriate actual proposition. Apparently, a proposition can be actual or observational without the agent's knowing that it is; and the agent can be mistaken in thinking a proposition actual or observational.

The point and meaning of the requirement that the effect be 'direct,' in the definitions of 'actual' and 'observational,' can be illustrated by considering the case of a sleeper who awakens and sees that the sun is shining. Then one might take the observation to have shown him, directly, that the sun is shining, and to have shown him indirectly that it is daytime. In general, an observation will cause numerous changes in the observer's belief function, but many of these can be construed as consequences of others. If there is a proposition E such that the *direct* effect of the observation is to change the observer's degree of belief in E to 1, then for any proposition A in the observer's preference ranking, his degree of belief in A after the observation will be the conditional probability

$$prob_E \, A \; = \; prob \, (A/E) \; = \; prob \, AE/prob \, E, \qquad (9)$$

where *prob* is the observer's belief function before the observation. And conversely, if the observer's belief function after the observation is $prob_E$ and $prob_E$ is not identical with *prob*, then the *direct* effect of the observation will be to change the observer's degree of belief in E to 1. This completes a definition of *direct*.

But from a certain strict point of view, it is rarely or never that there is a proposition for which the direct effect of an observation is to change the observer's degree of belief in that proposition to 1; and from that point of view, the classes of propositions that count as observational or actual in the sense defined above are either empty or as good as empty for practical purposes. For if we care seriously to distinguish between 0.999 999 and 1.000 000 as degrees of belief, we may find that, after looking out the window, the observer's degree of belief in the proposition that the sun is shining is not quite 1, perhaps because he thinks there is one chance in a million that he is deluded or deceived in some way; and similarly for acts where we can generally take ourselves to be at best *trying* (perhaps with very high probability of success) to make a certain proposition true.

One way in which philosophers have tried to resolve this difficulty is to postulate a phenomenalistic language in which an appropriate proposition E can always be expressed, as a report on the immediate content of experience;

(6)   G. E. M. Anscombe, *Intention,* § 8, Oxford, 1957; 2nd ed., Ithaca, N.Y., 1963.

but for excellent reasons, this move is now in low repute.[7] The crucial point is not that 0.999 999 is so close to 1.000 00 as to make no odds, practically speaking, for situations abound in which the gap is more like one half than one millionth. Thus, in examining a piece of cloth by candlelight one might come to attribute probabilities 0.6 and 0.4 to the propositions G that the cloth is green and B that it is blue, without there being any proposition E for which the direct effect of the observation is anything near changing the observer's degree of belief in E to 1. One might think of some such proposition as that (E) *the cloth looks green or possibly blue,* but this is far too vague to yield *prob* $(G/E) = 0.6$ and *prob* $(B/E) = 0.4$. Certainly, there is *something* about what the observer sees that leads him to have the indicated degrees of belief in G and in B, but there is no reason to think the observer can express this something by a statement in his language. And physicalistically, there is some perfectly definite pattern of stimulation of the rods and cones of the observer's retina which promotes his belief, but there is no reason to expect him to be able to describe that pattern or to recognize a true description of it, should it be suggested.

As Austin[8] points out, the crucial mistake is to speak seriously of the *evidence* of the senses. Indeed the relevant experiences have perfectly definite characteristics by virtue of which the observer comes to believe as he does, and by virtue of which in our example he comes to have degree of belief 0.6 in G. But it does not follow that there is a proposition E of which the observer is certain after the observation and for which we have *prob* $(G/E) = 0.6$, *prob* $(B/E) = 0.4$, etc.

In part, the quest for such phenomenological certainty seems to have been prompted by an inability to see how uncertain evidence can be used. Thus C. I. Lewis:

> If anything is to be probable, then something must be certain. The data which themselves support a genuine probability must themselves be certainties. We do have such absolute certainties, in the sense data initiating belief and in those passages of experience which later may confirm it. But neither such initial data nor such later verifying passages of experience can be phrased in the language of objective statement—because what can be so phrased is never more than probable. Our sense certainties can only be formulated by the expressive use of language, in which what is signified is a content of experience and what is asserted is the givenness of this content.[9]

(7)   See, e.g., J. L. Austin, *Sense and Sensibilia,* Oxford, 1962.
(8)   Austin, *op. cit.,* ch. 10.
(9)   C. I. Lewis, *An Analysis of Knowledge and Valuation,* La Salle, Illinois, 1946, p. 186.

But this motive for the quest is easily disposed of.[10] Thus, in the example of observation by candlelight, we may take the direct result of the observation (in a modified sense of 'direct') to be, that the observer's degrees of belief in G and B change to 0.6 and 0.4. Then his degree of belief in any proposition A in his preference ranking will change from *prob A* to

$$PROB\ A\ =\ 0.6\ prob\ (A/G) + 0.4\ prob\ (A/B).$$

In general, suppose that there are propositions $E_1, E_2, \ldots, E_n$, in which the observer's degrees of belief after the observation are $p_1, p_2, \ldots, p_n$; where the E's are pairwise incompatible and collectively exhaustive; where for each i, *prob* E_i is neither 0 nor 1; and where for each proposition A in the preference ranking and for each i the conditional probability of A on E_i is unaffected by the observation:

$$PROB\ (A/E_i)\ =\ prob\ (A/E_i). \tag{10}$$

Then the belief function after the observation may be taken to be *PROB*, where

$$PROB\ A\ =\ p_1 prob\ (A/E_1) + p_2 prob\ (A/E_2) + \cdots + p_n prob\ (A/E_n), \tag{11}$$

if the observer's preference rankings before and after the observation are both coherent. Where these conditions are met, the propositions $E_1, E_2, \ldots, E_n$, may be said to form a *basis* for the observation; and the notion of a basis will play the role vacated by the notion of *directness*.

The situation is similar in the case of acts. A marksman may have a fairly definite idea of his chances of hitting a distant target, e.g. he may have degree of belief 0.3 in the proposition H that he will hit it. The basis for this belief may be his impressions of wind conditions, quality of the rifle, etc.; but there need be no reason to suppose that the marksman can express the relevant data; nor need there be any proposition E in his preference ranking in which the marksman's degree of belief changes to 1 upon deciding to fire at the target, and for which we have *prob* $(H/E) = 0.3$. But the pair H, $\overline{H}$ may constitute a *basis* for the act, in the sense that for any proposition A in the marksman's preference ranking, his degree of belief after his decison is

$$PROB\ A\ =\ 0.3\ prob\ (A/H) + 0.7\ prob\ (A/\overline{H}).$$

It is correct to describe the marksman as *trying* to hit the target; but the

(10)   Jeffrey, *op. cit.*, ch. 11.

proposition that he is trying to hit the target can not play the role of E above. Similarly, it was correct to describe the cloth as *looking* green or possibly blue; but the proposition that the cloth looks green or possibly blue does not satisfy the conditions for directness.

The notion of directness is useful as well for the resolution of unphilosophical posers about probabilities, in which the puzzling element sometimes consists in failure to think of an appropriate proposition E such that the direct effect of an observation is to change degree of belief in E to 1, e.g. in the following problem reported by Mosteller.[11]

> Three prisoners, *a*, *b*, and *c*, with apparently equally good records have applied for parole. The parole board has decided to release two of the three, and the prisoners know this but not which two. A warder friend of prisoner *a* knows who are to be released. Prisoner *a* realizes that it would be unethical to ask the warder if he, *a*, is to be released, but thinks of asking for the name of *one* prisoner *other than himself* who is to be released. He thinks that before he asks, his chances of release are $\frac{2}{3}$. He thinks that if the warder says '*b* will be released,' his own chances have now gone down to $\frac{1}{2}$, because either *a* and *b* or *b* and *c* are to be released. And so *a* decides not to reduce his chances by asking. However, *a* is mistaken in his calculations. Explain.

Here indeed the possible cases (in a self-explanatory notation) are

$$AB, AC, BC,$$

and these are viewed by *a* as equiprobable. Then *prob A* is $\frac{2}{3}$ but *prob* (A/B) = *prob* $(A/C) = \frac{1}{2}$, and, since the warder must answer either '*b*' or '*c*' to *a*'s question it looks as if the direct result of the 'observation' will be that *a* comes to attribute probability 1 either to the proposition B that *b* will be released, or to the proposition C that *c* will be released. But this is incorrect. The relevant evidence-proposition would be more like the proposition *that the warder says, 'b,' or that the warder says, 'c,'* even though neither of these will quite do. For it is only in cases AB and AC that the warder's reply is dictated by the facts: in case BC, where *b* and *c* are both to be released, the warder must somehow choose *one* of the two true answers. If *a* expects the warder to make the choice by some such random device as tossing a coin, then we have *prob* $(A/$the warder says, '*b*') = *prob* $(A/$the warder says, '*c*') = *prob A* = $\frac{2}{3}$; while if *a* is sure that the warder will say '*b*' if he can, we have *prob* $(A/$the warder says '*b*') = $\frac{1}{2}$ but *prob* $(A/$the warder says '*c*') = 1.

---

(11)   Problem 13 of Frederick Mosteller, *Fifty Challenging Problems in Probability*, Reading, Mass., Palo Alto, and London, 1965.

3. BELIEF: REASONS VS. CAUSES

Indeed it is desirable, where possible, to incorporate the results of observation into the structure of one's beliefs via a basis of form $E, \overline{E}$ where the probability of E after the observation is nearly 1. For practical purposes, E then satisfies the conditions of directness, and the 'direct' effect of the observation can be described as informing the observer of the truth of E. Where this is possible, the relevant passage of sense experience *causes* the observer to believe E; and if *prob* (A/E) is high, his belief in E may be a *reason* for his believing A, and E may be spoken of as (inconclusive) *evidence* for A. But the sense experience is evidence neither for E nor for A. Nor does the situation change when we speak physicalistically in terms of patterns of irritation of our sensory surfaces, instead of in terms of sense experience: such patterns of irritation *cause* us to believe various propositions to various degrees; and sometimes the situation can be helpfully analyzed into one in which we are caused to believe $E_1, E_2, \ldots, E_n$, to degrees $p_1, p_2, \ldots, p_n$, whereupon those beliefs provide *reasons* for believing other propositions to other degrees. But patterns of irritation of our sensory surfaces are not reasons or evidence for any of our beliefs, any more than irritation of the mucous membrane of the nose is a *reason* for sneezing.

When I stand blinking in bright sunlight, I can no more believe that the hour is midnight than I can fly. My degree of belief in the proposition that the sun is shining has two distinct characteristics. (a) It is 1, as close as makes no odds. (b) It is compulsory. Here I want to emphasize the second characteristic, which is most often found in conjunction with the first, but not always. Thus, if I examine a normal coin at great length, and experiment with it at length, my degree of belief in the proposition that the next toss will yield a head will have two characteristics. (a) It is $\frac{1}{2}$. (b) It is compulsory. In the case of the coin as in the case of the sun, I cannot decide to have a different degree of belief in the proposition, any more than I can decide to walk on air.

In my scientific and practical undertakings I must make use of such compulsory beliefs. In attempting to understand or to affect the world, I cannot escape the fact that I am part of it: I must rather make use of that fact as best I can. Now where epistemologists have spoken of observation as a source of *knowledge*, I want to speak of observation as a source of compulsory *belief* to one or another degree. I do not propose to identify a very high degree of belief with knowledge, any more than I propose to identify the property of being near 1 with the property of being compulsory.

Nor do I postulate any *general* positive or negative connection between the characteristic of being compulsory and the characteristic of being sound or appropriate in the light of the believer's experience. Nor, finally, do I take a compulsory belief to be necessarily a permanent one: new experience or new reflection (perhaps, prompted by the arguments of others) may loosen the bonds of compulsion, and may then establish new bonds; and the effect may be that the new state of belief is sounder than the old, or less sound.

Then why should we trust our beliefs? According to K. R. Popper,

> ... the decision to accept a basic statement, and to be satisfied with it, is causally connected with our experiences—especially with our *perceptual experiences*. But we do not attempt to *justify* basic statements by these experiences. Experiences can *motivate a decision,* and hence an acceptance or a rejection of a statement, but a basic statement cannot be *justified* by them—no more than by thumping the table.[12]

I take this objection to be defective, principally in attempting to deal with basic statements (observation reports) in terms of *decisions* to *accept* or to *reject* them. Here acceptance parallels belief, rejection parallels disbelief (belief in the denial), and tentativeness or reversibility of the decision parallels *degree* of belief. Because logical relations hold between statements, but not between events and statements, the relationship between a perceptual experience (an event of a certain sort) and a basic statement cannot be a logical one, and therefore, Popper believes, cannot be of a sort that would justify the statement:

> Basic statements are accepted as the result of a decision or agreement; and to that extent they are conventions.[13]

But in the absence of a positive account of the nature of acceptance and rejection, parallel to the account of partial belief given in section 1, it is impossible to evaluate this view. Acceptance and rejection are apparently acts undertaken as results of decisions; but somehow the decisions are conventional—perhaps only in the sense that they may be *motivated* by experience, but not *adequately* motivated, if adequacy entails justification.

To return to the question, 'Why should we trust our beliefs?' one must ask what would be involved in *not* trusting one's beliefs, if belief is analyzed as in section 1 in terms of one's preference structure. One way of mistrusting a belief is declining to act on it, but this appears to consist merely in lowering the degree of that belief: to mistrust a partial belief is then to alter its degree to a new, more suitable value.

A more hopeful analysis of such mistrust might introduce the notion of sensitivity to further evidence or experience. Thus, agents 1 and 2 might have the same degree of belief $-\frac{1}{2}$ — in the proposition H_1 that the first toss of a

(12)   K. R. Popper, *The Logic of Scientific Discovery,* London, 1959, p. 105.
(13)   Popper, *op. cit.,* p. 106.

certain coin will yield a head, but agent 1 might have this degree of belief because he is convinced that the coin is normal, while agent 2 is convinced that it is either two-headed or two-tailed, he knows not which.[14] There is no question here of agent 2's expressing his mistrust of the figure $\frac{1}{2}$ by lowering or raising it, but he can express that mistrust quite handily by aspects of his belief function. Thus, if H_i is the proposition that the coin lands head up the ith time it is tossed, agent 2's beliefs about the coin are accurately expressed by the function $prob_2$ where

$$prob_2\, H_i = \tfrac{1}{2}, \; prob_2\, (H_i/H_j) = 1,$$

while agent 1's beliefs are equally accurately expressed by the function $prob_1$ where

$$prob_1\, (H_{i_1}, H_{i_2}, \ldots, H_{i_n}) = 2^{-n},$$

if $i_1 < i_2 < \cdots < i_n$. In an obvious sense, agent 1's beliefs are *firm* in the sense that he will not change them in the light of further evidence, since we have

$$prob_1\, (H_{n+1}/H_1, H_2, \ldots, H_n) = prob_1\, H_{n+1} = \tfrac{1}{2},$$

while agent 2's beliefs are quite tentative and in that sense, mistrusted by their holder. Still, $prob_1\, H_i = prob_2\, H_i = \tfrac{1}{2}$.

After these defensive remarks, let me say how and why I take compulsive belief to be sound, under appropriate circumstances. Bemused with syntax, the early logical positivists were chary of the notion of truth; and then, bemused with Tarski's account of truth, analytic philosophers neglected to inquire how we come to believe or disbelieve simple propositions. Quite simply put, the point is: coming to have suitable degrees of belief in response to experience is a matter of training—a *skill* which we begin acquiring in early childhood, and are never quite done polishing. The skill consists not only in coming to have appropriate degrees of belief in appropriate propositions under paradigmatically good conditions of observation, but also in coming to have appropriate degrees of belief between zero and one when conditions are less than ideal.

Thus, in learning to use English color words correctly, a child not only learns to acquire degree of belief 1 in the proposition that the cloth is blue, when in bright sunlight he observes a piece of cloth of uniform hue, the hue

(14) This is a simplified version of 'the paradox of ideal evidence,' Popper, *op. cit.*, pp. 407-409.

being squarely in the middle of the blue interval of the color spectrum: he also learns to acquire appropriate degrees of belief between 0 and 1 in response to observation under bad lighting conditions, and when the hue is near one or the other end of the blue region. Furthermore, his understanding of the English color words will not be complete until he understands, in effect, that blue is between green and violet in the color spectrum: his understanding of this point or his lack of it will be evinced in the sorts of mistakes he does and does not make, e.g. in mistaking green for violet he may be evincing confusion between the meanings of 'blue' and of 'violet,' in the sense that his mistake is linguistic, not perceptual.

Clearly, the borderline between factual and linguistic error becomes cloudy, here: but cloudy in a perfectly realistic way, corresponding to the intimate connection between the ways in which we experience the world and the ways in which we speak. It is for this sort of reason that having the right language can be as important as (and can be in part identical with) having the right theory.

Then learning to use a language properly is in large part like learning such skills as riding bicycles and flying aeroplanes. One must train oneself to have the right sorts of responses to various sorts of experiences, where the responses are degrees of belief in propositions. This may, but need not, show itself in willingness to utter or assent to corresponding sentences. Need not, because e.g. my cat is quite capable of showing that it thinks it is about to be fed, just as it is capable of showing what its preference ranking is, for hamburger, tuna fish, and oat meal, without saying or understanding a word. With people as with cats, evidence for belief and preference is behavioral; and speech is far from exhausting behavior.[15]

Our degrees of beliefs in various propositions are determined jointly by our training and our experience, in complicated ways that I cannot hope to describe. And similarly for conditional subjective probabilities, which are certain ratios of degrees of belief: to some extent, these are what they are because of our training—because we speak the languages we speak. And to this extent, conditional subjective probabilities reflect *meanings*. And in this sense, there can be a theory of degree of confirmation which is based on analysis of meanings of sentences. Confirmation theory is therefore semantical and, if you like, logical.

(15)   Jeffrey, *op. cit.*, pp. 57-59.

Selected Bibliography

The present bibliography makes no pretense to listing in a fairly complete way those articles and books relevant to the subject of subjective probability but it is hoped that the selection is reasonably representative. A number of standard books and articles which state alternative points of view have been listed; in addition works which are cited in the introduction have been included here.

Adams, Ernest W.
[1960]   "Survey of Bernouillian Utility Theory," *Mathematical Thinking in the Measurement of Behavior,* Solomon (ed.) (1960), 151-268.
[1964]   "On Rational Betting Systems," *Archiv für Mathematische Logik und Grundlagenforschung,* 6/1-2 (1964), 7-29; 6/3-4 (1964), 112-128.
[1976]   "Prior Probabilities and Counterfactuals," in Harper, W. and Hooker, C. (eds.) *Foundations of Probability Theory, Statistical Inference and Statistical Theories of Science,* 3 vols., Reidel Publishing Co., Dordrecht, 1976, v. 1, 1-22.

Agassi, Joseph
[1975]   "Subjectivism: From Infantile Disease to Chronic Illness," *Synthese,* **30** (1975), 3-14.

Allais, Maurice
[1953]   "Le comportement de l'homme rationel devant la risque: critique des postulats de l'école americaine," *Econometrica,* **21** (1953), 503-546.

Anscombe, F. J.
[1961]   "Bayesian Statistics," *American Statistician,* **15** (1961), 21-24.

Anscombe, F. J.; Aumann, R. J.
[1963]   "A Definition of Subjective Probability," *Annals of Mathematical Statistics,* **34** (1963), 199-205.

Arrow, Kenneth J.
[1952]   "Alternative Approaches to the Theory of Choice in Risk-Taking Situations," *Econometrica,* **19** (1951), 404-437. (Reprinted: *Cowles Commission Paper No. 51,* Chicago, 1952.)
[1958]   "Utilities, Attitudes, Choices; A Review Article," *Econometrica,* **26** (1958), 1-28.
[1963]   *Social Choice and Individual Values,* Second Edition, Wiley & Sons, New York, 1963. (First Edition, 1951).
[1966]   Exposition of the Theory of Choice Under Uncertainty," *Synthese,* **16** (1966), 253-269.

Baillie, P.
[1973]   "Confirmation and the Dutch Book Argument," *British Journal for the Philosophy of Science,* **24** (1973), 393-397.

Barnard, George A.
[1947]   "The Meaning of a Significance Level," *Biometrika,* **34** (1947), 179-182.
[1962]   "Comments on Savage," in Barnard and Cox (eds.), *Foundations of Statistical Inference,* 1962, 39-49.

Barnard, George A. and Cox, D. R.
[1962]   *The Foundations of Statistical Inference,* Wiley and Sons, New York, 1962.

Barnard, G. A.; Jenkins, G. M.; Winsten, C. B.
[1962]   "Likelihood, Inference, and Time Series," *Journal of the Royal Statistical Society, Series A,* **125** (1962), 321-372.

Bartlett, Maurice S.
[1940]   "The Present Position of Mathematical Statistics," *Journal of the Royal Statistical Society,* **103** (1940), 1-29.
[1949]   "Probability in Logic, Mathematics, and Science," *Dialectica,* **3** (1949), 104-113.
1962a]   *Essays in Probability and Statistics,* Methuen, London, 1962.
[1962b] "Comments on Savage," in Barnard and Cox, eds., *Foundations of Statistical Inference,* 1962, 36-39.

Bayes, Thomas
[1940]   *Facsimile of Two Paper of Bayes, i. An Essay towards Solving a Problem in the Doctrine of Chances, with Richard Price's Foreword and Discussions; Philosophical Transactions of the Royal Society, pp. 370-418, 1763. With a Commentary by Edward C. Monina: ii. A Letter on Asymptotic Series from Bayes to John Canton, pp. 269-271 of the same volume. With a commentary by W. Edward Demming. (Edited by W. Edward Demming.)* The Graduate School, Department of Agriculture, Washington, D.C., 1940. [(*i*) has been reprinted with a biographical note by G. A. Barnard, *Biometrika,* **45** (1958), 293-315.]

Bernoulli, Daniel
[1738]   "Specium Theoriae Novae De Mensura Sortis," *Commentarii Academiae Scientarum Imperialis Petropolitanae* (for 1730 and 1731), **5** (1738), 175-192.
[1954]   "Exposition of a New Theory on the Measurability of Wealth" (English Translation of [B1738] by Louise Sommer), *Econometrica,* **22** (1954), 23-36.

Bernoulli, Jacob
[1713]   *Ars Conjectandi,* Basel, 1713.

Birnbaum, Allan
[1962]   "On the Foundations of Statistical Inference," *Journal of the American Statistical Association,* **47** (1962), 269-306.

Blackwell, D. and Dubins, L.
[1962]   "Merging of Opinions with Increasing Information," *Annals of Mathematical Statistics,* **33** (1962), 882-886.

Blackwell, D. and Girshick, M. A.
[1954]   *Theory of Games and Statistical Decisions,* Wiley and Sons, New York, 1954.

Blum, J. R. and Rosenblatt, J.
[1967]   "On Partial a priori Information in Statistical Inference," *Annals of Mathematical Statistics,* **38** (1967), 1671-1678.

Bolker, Ethan
[1967]   "A Simultaneous Axiomatization of Utility and 'Subjective Probability,'"
         Philosophy of Science, **34** (1967), 333-340.

Borel, Emile
[1924]   "A propos d'un traite de probabilités," *Revue Philosophique*, **98** (1924),
         321-336.
[1939]   *Valeur pratique et philosophie des probabilités*, Gauthier-Villars, Paris, 1939.
[1943]   *Les probabilités et la vie*, Presses Universitaires de France, Paris, 1943.
[1962]   *Probabilities and Life* (English translation of [1943] by Maurice Baudin),
         Dover, New York, 1962.
[1964]   "Apropos of a Treatise on Probability," in *Studies in Subjective Probability*,
         Wiley and Sons, New York, 1964, First Edition, Kyburg and Smokler (eds.).

Braithwaite, Richard Bevan
[1953]   *Scientific Explanation*, Cambridge University Press, Cambridge, 1953
         (Reprinted as Harper Torchbook, New York, 1960).
[1962]   "On Unknown Probabilities," *Observation and Interpretation in the Philoso-
         phy of Physics* (Ed. Stephen Korner), Dover, New York, 1962, 3-11.

Brown, G. Spencer
[1957]   *Probability and Scientific Inference*, Longmans, London, 1957.

Bush, R. and Mosteller, F.
[1955]   *Stochastic Models for Learning*, Wiley and Sons, New York, 1955.

Carnap, Rudolf
[1944]   "The Two Concepts of Probability," *Philosophy and Phenomenological
         Research*, **5** (1944-45), 513-532.
[1945]   "On Inductive Logic," *Philosophy of Science*, **12** (1945), 72-97.
[1947]   "On the Application of Inductive Logic," *Philosophy and Phenomenological
         Research*, **8** (1947-48), 133-148.
[1952]   *The Continuum of Inductive Methods*, University of Chicago Press, Chicago,
         1962.
[1953]   "What is Probability?" *Scientific American*, **189** (1953), 128-138.
[1962a] *Logical Foundations of Probability*, Second Edition, University of Chicago
         Press, Chicago, 1962 (First Edition, 1950).
[1962b] "The Aim of Inductive Logic," *Logic, Methodology and Philosophy of Science*
         (ed. Nagel, Suppes, Tarski), Stanford University Press, Stanford, 1962,
         303-318.
[1968]   "Inductive Logic and Inductive Intuition," *The Problem of Inductive Logic*,
         Lakatos (ed.), 1968, 258-267.
[1971a] "Inductive Logic and Rational Decisions," *Studies in Inductive Logic and
         Probability I*, Carnap and Jeffrey (eds.), 1971, 5-31.
[1971b] "A Basic System of Inductive Logic, Part I," *Studies in Inductive Logic and
         Probability I*, Carnap and Jeffrey (eds.), 1971, 33-165.

Carnap, R. and Jeffrey, R.
[1971]   *Studies in Inductive Logic and Probability I*, University of California Press,
         Berkeley, Los Angeles, and London, 1971.

Cohen, John and Hansel, M.
[1956]   *Risk and Gambling: the Study of Subjective Probability*, Philosophical
         Library, New York, 1956.

Cox, D. R.
[1962]   "Comments on Savage," *Foundations of Statistical Inference,* Barnard (ed.),
         1962, 49-53.

Cox, D. R. and Barnard, George: *See* Barnard, George and Cox, D. R.

Cox, R. T.
[1946]   "Probability Frequency and Reasonable Expectation," *American Journal of
         Physics,* 14 (1946), 1-13.

Churchman, C. and Ratoosh, P.
[1959]   *Measurement: Defintion and Theories,* Wiley and Sons, New York, 1959.

Cramer, Harold
[1951]   *Mathematical Methods of Statistics,* Princeton University Press, Princeton,
         1951.

Dantzig, D. van
[1957]   "Statistical Priesthood (Savage on Personal Probabilities)," *Statistica
         Neerlandica,* 2 (1957), 1-16.

Davidson, Barbara and Ellis, Brian
[1976]   "Logic and Strict Coherence," *Reports on Mathematical Logic, No. 6,* The
         Jagieulluman University of Cracow; The Silesian University of Katowice,
         Warsaw, 1976, 29-40.

Davidson, Donald
[1974]   "Belief and the Basis of Meaning," *Synthese,* 27 (1974), 309-323.
[1976]   "Hempel on Explaining Actions," *Erkenntnis,* 10 (1976), 239-253.

Davidson, Donald and Suppes, Patrick
[1956]   "A Finite Axiomatization of Subjective Probability and Utility," *Econo-
         metrica,* 24 (1956), 264-275.

Davidson, Donald; Suppes, Patrick; Siegel, Sidney
[1957]   *Decision Making: An Experimental Approach,* Stanford University Press,
         Stanford, 1957.

De Finetti: *See* Finetti, Bruno de.

De Morgan, Augustus
[1847]   *Formal Logic,* Taylor and Walton, London, 1847.

Dempster, Arthur P.
[1962]   "A Subjectivist Look at Robustness," *Bulletin of the International Statistical
         Institute,* 46, Book I, 349-374.
[1966]   "New Methods for Reasoning Towards Posterior Distributions Based on
         Sample Data," *Annals of Mathematical Statistics,* 37 (1966), 355-374.
[1967a] "Upper and Lower Probability Inferences Based on a Sample from a Finite
         Univariate Population," *Biometrika,* 54 (1967), 515-528.
[1967b] "Upper and Lower Probabilities induced by a Multivalued Mapping," *Annals
         of Mathematical Statistics,* 38 (1967), 325-339.
[1968]   "A Generalization of Bayesian Inference," *Journal of the Royal Statistical
         Society, Series B,* 30 (1968), 205-247.

Diaconos, Persi
[1977]   "Finite Forms of de Finetti's Theorem of Exchangeability," *Synthese,* 36
         (1977), 271-281.

Diaconos, Persi and Freedman, David
[1978]   "De Finetti's Generalization of Exchangeability," Technical Report, no. 109
         (March 1978), Department of Statistics, Stanford University.

Domotor, Zoltan
[1978]   "Axiomatization of Jeffrey Utilities," *Synthese,* **38** (1978), 165-210.

Dorling, Jon
[1972]   "Bayesianism and the Rationality of Scientific Inference," *British Journal
         For the Philosophy of Science,* **23** (1972), 181-190.

Dubins, L. and Blackwell, D.: *See* Blackwell, D. and Dubins, L.

Edwards, Ward
[1954]   "The Theory of Decision Making," *Psychological Bulletin,* **51** (1954), 380-
         417.
[1960]   "Measurement of Utility and Subjective Probability," in Gulliksen & Messick
         (eds.), *Psychological Scaling,* 1960.
[1961]   "Behavioral Decision Theory," *Annual Review of Psychology,* **12** (ed.
         Farnsworth), Annual Reviews, Palo Alto, 1961, 473-498.
[1962]   "Subjective Probabilities Inferred from Decisions," *Psychological Review,* **69**
         (1962), 109-135.

Edwards, Ward; Lindman, Harold; Savage, Leonard J.
[1963]   "Bayesian Statistical Inference for Psychological Research," *Psychological
         Review,* **70** (1963), 193-242.

Eisenberg, Edward and Gale, David
[1959]   "Consensus of Subjective Probabilities: The Pari-Mutuel Method," *Annals
         of Mathematical Statistics,* **30** (1959), 165-168.

Ellis, Brian
[1973]   "The Logic of Subjective Probability," *British Journal for the Philosophy of
         Science,* **24** (1973), 125-152.
[1979]   *Rational Belief Systems,* Blackwell, Oxford, 1979.

Ellis, Brian and Davidson, Barbara: *See* Davidson, Barbara and Ellis, Brian

Ellsberg, Daniel
[1963]   "Risk, Ambiguity, and the Savage Axioms; A Reply," *Quarterly Journal of
         Economics,* **LXXVII** (1963), 336-342.

Field, Hartry
[1977]   "Logic, Meaning, and Conceptual Role, *Journal of Philosophy,* **LXXIV**
         (1977), 379-408.
[1978]   "A Note on Jeffrey Conditionalization," *Philosophy of Science,* **45** (1978),
         361-367.

Fellner, William
[1961]   "Distribution of Subjective Probabilities as a Response to Uncertainty,"
         Quarterly Journal of Economics, **LXXV** (1961), 670-689.
[1965]   *Probability and Profit,* E. D. Irwin, Homewood, Illinois, 1965.

Fine, Terrence, L.
[1973]   *Theories of Probability,* Academic Press, New York, 1973.

Finetti, Bruno de

[1931]   "Sul significato doggetivo della probabilità," *Fundamenta Mathematical,* **17** (1931), 298-329.

[1937]   "La Prevision: ses lois logiques, ses sources subjectives," *Annales de l'Institut Henri Poincaré,* 7 (1937), 1-68.

[1938]   "Sur la condition d'équivalence partielle," Vol. 6 of *Colloque consacré à la théorie des probabilités,* Hermann et Cie, Paris, 1938, *Act. Scient.* **739**, 5-18.

[1949a] "Le vrai et la probable," *Dialectica,* **3** (1949), 78-92.

[1949b] "Sulla impostazione assiomatica del calcolo delle probabilità," *Annali Triestini,* series 2, **19** (1949), 29-81.

[1949c] "La 'logica del plausible' secondo la concezione di Polya," Vol. 1 of *Atti della XLII Riunione della Società Italiana per il Progresso delle Scienze,* Società Italiana per il Progresso delle Scienze, Rome, 1949, 227-236.

[1951a] "Recent Suggestions for the Reconciliation of Theories of Probability," *Proceedings of the Second Berkeley Symposium on Mathematical Statistics and Probability* (ed. Jerzy Neyman), University of California Press, Berkeley, 1951, 217-226.

[1951b] "Aggiunto alla nota assiomatica della probabilità," *Annali Triestini,* series 2, **20** (1951), 5-22.

[1951c] "Role et domaine d'application du théorème de Bayes selon les differents points de vue sur les probabilités," Congrès International de Philosophie des Sciences, Paris, 1949, Vol. IV (*Calcul des probabilités*), Hermann et Cie, Paris, 1951, *Act. Sient. Ind.,* **1146**, 49-66.

[1952a] "Sulla preferabilità," *Giornale degli economisti e annali di economia,* **11** (1952), 685-709.

[1952b] "La notion de 'distribution d'opinions' comme case d'un essai d'interpretation de la statistique," *Publications de l'institut de statistique de l'université de Paris,* **1** (1952), 1-11.

[1954]   "Concetti sul comportomento induttivo illustrati su di un essempio," *Statistica,* **14** (154), 350-378.

[1955a] "Les problèmes psychologiques sur les probabilités subjectives," *Journal Psych. Normale et Path.,* **52** (1955), 253-259.

[1955b] "Experience et théorie dans l'application d'une domaine scientifique," *Revue de metaphysique et du morale,* **60** (1955), 264-285.

[1958]   "Foundations of Probability," Vol. 1 of *Philosophy in the Mid-Century* (ed. R. Klibansky), La Nuova Italia Editrice, Florence, 1958, 140-147.

[1959]   "La probabilità e la statistica nei rapporti con l'induzione secondo i diversi punti di vista," *Induzione e statistica,* Istituto Matematico dell'Università, Cremonese, Rome, 1959, 115 pp.

[1961]   "Dans quel sens la théorie des décisions est-elle et doit-elle être normative?" *La décision (colloques internationaux du Centre National de la Recherche),* May, 1958. Editions du C.N.R.S., Paris, 1961, 159-169.

[1962]   "Does It Make Sense to Speak of 'Good Probability Appraisers'?" *The Scientist Speculates* (ed. I. J. Good), Heinemann, London, 1962, 357-364.

[1964]   "Foresight: Its Logical Laws, its Subjective Sources," in Kyburg and Smokler, *Studies in Subjective Probability,* Wiley and Sons, New York, 1964, 93-158.

[1969]   "Initial Probabilities: A Prerequisite for any Valid Induction," *Synthese,* **20** (1969), 2-16.

[1972]   *Probability, Induction and Statistics,* Wiley and Sons, New York, 1972.

Finetti, Bruno de
[1973]   "Bayesianism: Its Unifying Role for Both the Foundations and the Applica-
          tions of Statistics," *Proceedings of the 39th Session of the International
          Statistical Institute,* 1973, Vienna, 349-368.
[1974-75] *Theory of Probability,* 2 vols., Wiley and Sons, New York, 1974-75, trans-
          lated from *Teoria delle Probabilità,* 2 vols., Einaudi, Torino, 1970.

Finetti, Bruno de and Savage, Leonard J.
[1962]   "Sul modo di scegliere le probabilità iniziali," *Sui fondamenti della statistica,*
          Biblioteca Del Metron, Series C, **1**, 1962, 81-147 (English summary, 148-151).

Fishburn, P. C.
[1967]   "Preference-Based Definitions of Subjective Probability," *Annals of Mathe-
          matical Statistics,* **38** (1967), 1605-1619.

Fisher, Roland A.
[1955]   "Statistical Methods and Scientific Induction," *Journal of the Royal Statisti-
          cal Society* (Series B), **17** (1955), 69-78.
[1956]   *Statistical Methods and Scientific Inference,* Oliver and Boyd, Edinburgh,
          1956.

Fortet, Robert M.
[1951]   "Faut-il élargir les axioms du calcul des probabilités?" Congrès de philosophie
          des sciences, Paris, 1949, Vol. IV (*Calcul des probabilités*), 35-47, Hermann
          et Cie, Paris, 1951, *Act. Sci. Ind.,* **1146**.

Fréchet, Maurice
[1954]   "Une problème psychologique sur les probabilités subjectives irrationelles,"
          Journal Psych. Normale et Path., **47-51** (1950-54), 431-438.
[1955]   "Sur l'importance en économétrie de la distinction entre les probabilités
          rationelles et irrationelles," *Econometrica,* **23** (1955), 303-306.

Freedman, David
[1962]   "Invariants under Mixing which Generalize de Finetti's Theorem," *Annals of
          Mathematical Statistics,* **33** (1962), 916-923.

Freedman, David and Diaconos, Persi: *See* Diaconos, Persi and Freedman, David

Friedman, Milton and Savage, Leonard J.
[1946]   "The Utility Analysis of Choice Involving Risk," *Journal of Political
          Economy,* **56** (1946), 279-304.

[1952]   "The Expected-Utility Hypotheses and the Measurability of Utility," *Journal
          of Political Economy,* **60** (1952), 463-474.

Gale, David: *See* Eisenberg, E. and Gale, D.

Giles, Robin
[1976]   "A Logic for Subjective Belief," in Harper, Willian and Hooker, Clifford, eds.
          *Foundations of Probability Theory, Statistical Inference, and Statistical
          Theories of Science,* 3 vols., Reidel Publishing Co., Dordrecht, 1976, v. 1,
          41-70.

Gillies, Donald A.
[1972]   "The Subjective Theory of Probability," *British Journal for the Philosophy of
          Science,* **23** (1972), 138-156.

Girschick, M. and Blackwell, D.: *See* Blackwell, D. and Girschick, M.

Godambe, V. P. and Sprott, D. A.
[1971] *Foundations of Statistical Inference,* Holt, Rinehart and Winston of Canada, Toronto and Montreal, 1971.

Good, I. J.
[1950] *Probability and the Weighing of Evidence,* Hafner, New York, 1950.
[1952] "Rational Decisions," *Journal of the Royal Statistical Society, (Series B),* **14** (1952), 107-114.
[1959a] "Kinds of Probability," *Science,* **129** (1959), 443-447.
[1959b] "Weight of Evidence, Corroboration, Explanatory Power, Information, and the Utility of Experiments," *Journal of the Royal Statistical Society (Series B),* **22** (1959), 319-331.
[1962] "Subjective Probability as the Measure of a Non-Measurable Set," *Logic, Methodology and Philosophy of Science* (ed. Nagel, Suppes, Tarski), Stanford University Press, Stanford, 1962, 319-329.
[1965] *The Estimation of Probabilities: An Essay on Modern Bayesian Methods,* M.I.T. Research Monograph No. 30, The M.I.T. Press, Cambridge, 1965.
[1968] "Corroboration, Explanation, Evolving Probabilities, Simplicity, and a Sharpened Razor," *British Journal for the Philosophy of Science,* **19** (1968), 123-143.
[1975] Explicativity, Corroboration and the Relative Odds of Hypotheses, *Synthese,* **30** (1975), 39-74.
[1976] "The Bayesian Influence, or How to Sweep Subjectivism Under the Carpet," *Foundations of Probability Theory, Statistical Inference, and Statistical Theories of Science,* Vol. II, Harper and Hooker, eds., 1976, 125-174.

Gulliksen, Harold and Messick, Samuel (eds.)
[1960] *Psychological Scaling,* Wiley and Sons, New York and London, 1960.

Hacking, Ian
[1965a] *Logic of Statistical Inference,* Cambridge University Press, Cambridge, 1965.
[1965b] Review of *Studies in Subjective Probability* (Kyburg and Smokler, eds.), *British Journal for the Philosophy of Science,* **16** (1965-66), 334-339.
[1967] "Slightly More Realistic Personal Probability," *Philosophy of Science,* **34** (1967), 311-325.
[1968] "On Falling Short of Strict Coherence," *Philosophy of Science,* **35** (1968), 284-286.
[1975] *The Emergence of Probability,* Cambridge University Press, Cambridge, 1975.

Hailperin, Theodore
[1976] *Boole's Logic and Probability,* North Holland, Amsterdam, New York, Oxford, 1976.

Hallden, Soren
[1966] "On Preference, Probability, and Learning," *Synthese,* **16** (1966), 307-320.

Harper, William L.
[1975] "Rational Belief Change, Popper Functions, and Counterfactuals," *Synthese,* **30** (1975), 221-262.
[1976] "Rational Belief Change, Popper Functions, and Counterfactuals," in Harper and Hooker (eds.), *Foundations of Probability Theory, Statistical Inference, and Statistical Theories of Science,* 2 vols., Reidel Publishing Co., Dordrecht, 1976, 73-115.

Harper, William
[1978] "Bayesian Learning Models with Revision of Evidence," *Philosophia,* 7 (1978), 357-367.

Harper, William L. and Hooker, Clifford (eds.)
[1976] *Foundations of Probability Theory, Statistical Inference, and Statistical Theories of Science,* 2 vols., Reidel Publishing Co., Dordrecht, 1976.

Hartley, H. O.
[1963] "In Dr. Bayes' Consulting Room," *American Statistician,* 17 (1963), 22-24.

Heilig, Klaus
[1978] "Carnap and de Finetti on Bets and the Probability of Singular Events: The Dutch Book Argument Reconsidered," *British Journal for the Philosophy of Science,* 29 (1978), 325-346.

Hermes, Hans
[1957] "Uber eine logische Begrundung der Wahrscheinlichkeitstheorie," *Mathematisch-Physikalische Semesterberichte,* 5 (1957), 214-224.
[1958] "Zur Einfacheitsprinzip in der Wahrscheinlichkeitsrechung," *Dialectica,* 12 (1958), 317-331.

Hesse, Mary
[1975] "Bayesian Methods and the Initial Probabilities of Theories," *Minnesota Studies in the Philosophy of Science,* v. VI, eds. Maxwell, G., and Anderson, R., Univ. of Minn. Press, Minneapolis, 1975, 50-105.

Hewitt, Edward and Savage, Leonard, J.
[1955] "Symmetric Measures on Cartesian Products," *Transactions of the American Mathematical Society,* 80 (1955), 470-501.

Hilpinen, Risto
[1966] "On Inductive Generalization in Monadic First-Order Logic with Identity," Hintikka and Suppes (eds.), *Aspects of Inductive Logic,* 1966, 133-154.

Hintikka, Jaakko
[1964] "Towards a Theory of Inductive Generalization," *Logic, Methodology and Philosophy of Science,* Yehoshua Bar-Hillel (ed.), North-Holland Publishing Co., Amsterdam, 1964, 274-288.
[1965] "On a Combined System of Inductive Logic," *Studia Logico-Mathematica et Philosophica in Honorem Rolf Nevanlinna, Acta Philosophica Fennica,* 18 (1965), 21-30.
[1966] "A Two-Dimensional Continuum of Inductive Methods," Hintikka and Suppes (eds.), *Aspects of Inductive Logic,* 1966, 113-132.
[1971] "Unknown Probabilities, Bayesianism and de Finetti's Representation Theorem," *PSA 1970,* Buck and Cohen (eds.), 1971, 325-341.

Hintikka, Jaako and Suppes, Patrick (eds.)
[1966] *Aspects of Inductive Logic,* North-Holland Publishing Co., Amsterdam, 1966.

Hodges, J. L. and Lehmann, E. L.
[1952] "The Use of Previous Experience in Reaching Statistical Decisions," *Annals of Mathematical Statistics,* 23 (1952), 396-407.

Holstein, C. Stael von
[1973] " The Concept of Probability in Psychological Experiments," *Logic, Methodology and Philosophy of Science IV,* Suppes, et al. (eds.), 1973, 451-466.

Jaynes, E. T.
[1958]   *Probability Theory in Science and Engineering: Colloquium Lectures in Pure and Applied Science,* No. 4, Feb. 1958. Dallas, Texas: Socony Mobil Oil Corporation, 1959, Lecture 5.
[1968]   "Prior Probabilities," *IEEE Transactions on Systems Science and Cybernetics,* **4**, 1968.

Jeffrey, Richard C.
[1956]   "Valuation and Acceptance of Scientific Hypotheses," *Philosophy of Science,* **23** (1956), 237-246.
[1964]   New Foundations for Bayesian Decision Theory," *Logic, Methodology and Philosophy of Science,* ed. by Yehoshua Bar-Hillel, North-Holland Publishing Co., Amsterdam, 1964, 289-300.
[1965]   *The Logic of Decision,* McGraw-Hill Publishing Co., New York, 1965.
[1966]   "Solving the Problem of Measurement," *Journal of Philosophy,* **58**, 1966, 400-401.
[1968a] "The Whole Truth," *Synthese,* **18** (1968), 24-27.
[1968b] Review of Isaac Levi, *Gambling with Truth, Journal of Philosophy,* **65** (1968), 313-322.
[1970]   "Dracula meets Wolfman: Acceptance versus Partial Belief," in *Induction, Acceptance and Partial Belief,* Swain, Marshall, ed., Reidel Publishing Co., Dordrecht, 1970, 157-185.
[1974]   "Preferences among Preferences," *Journal of Philosophy,* **71** (1974), 377-391.
[1975a] "Carnap's Empiricism," *Minnesota Studies in the Philosophy of Science,* v. VI, ed. Maxwell, G. and Anderson, R. University of Minnesota Press, Minneapolis, 1975, 37-49.
[1975b] "Probability and Falsification; Critique of the Popper Program," *Synthese,* **30** (1975), 95-118.
[1977]   "A Note on the Kinematics of Preference," *Erkenntnis,* **11** (1977), 135-141.

Jeffrey, Richard C. and Carnap, Rudolph: *See* Carnap, R. and Jeffrey R. C.

Jeffreys, Harold
[1955]   "The Present Position of the Theory of Probability," *British Journal for the Philosophy of Science,* **5** (1955), 275-289.
[1957]   *Scientific Inference,* Second Edition, Cambridge, 1957 (First Edition, 1931).
[1961]   *Theory of Probability,* Third Edition, Oxford University Press, Oxford, 1961 (Second Edition, 1948; First Edition, 1938).

Jeffreys, Harold and Wrinch, Dorothy
[1919]   "On Some Aspects of the Theory of Probability," *The Philosophical Magazine Series 6,* **38** (1919), 715-731.

Jenkins, G. M.: *See* Barnard, G. A.; Jenkins, G. M.; Winsten, C. B.

Jones, Robert M.
[1965]   "The Non-Reducibility of Koopman's Theory of Probability in Carnap's System for M C," *Philosophy of Science,* **32** (1965), 368-369.

Kahneman, D. and Tversky, A.
[1974a] "Judgment under Uncertainty: Heuristics and Biases," *Science,* **185** (1974), 1124-1131.
[1974b] "Subjective Probability: A Judgment of Representativeness," in *The Concept of Probability in Psychological Experiments,* ed. C. A. S. Stael-von Holstein, 1974, 25-48.

Kemeny, John
[1955]   "Fair Bets and Inductive Probabilities," *Journal of Symbolic Logic,* **20** (1955), 263-273.

Kempthorne, Oscar
[1976]   "Statistics and the Philosophers," Harper and Hooker (eds.), *Foundations of Probability Theory, Statistical Inference and Statistical Theories of Science,* Vol. II, 1976, 273-314.

Keynes, John Maynard
[1921]   *A Treatise on Probability,* Macmillan, London, 1921 (Reprinted: Harper Torchbooks, New York, 1962).
[1933]   "F. P. Ramsey," *Essays in Biography,* Macmillan, London, 1933, 299-311.

Khinchin, A.
[1932]   "Sur les classes d'événements équivalents," *Mathematičeskii Sbornik (Recueil Math. Moscou),* **39** (1932), 40-43.
[1952]   "On Classes of Equivalent Events" (in Russian), *Doklady Akademii Nauk SSR N. S.,* **85** (1952), 713-714.

Koopman, Bernard Osgood
[1940a] "The Bases of Probability," *Bulletin of the American Mathematical Society,* **46** (1940), 763-774.
[1940b] "The Axioms and Algebra of Intuitive Probability," *Annals of Mathematics,* **41** (1940), 269-292.
[1941]   "Intuitive Probabilities and Sequences," *Annals of Mathematics,* **42** (1941), 169-187.

Kraft, Charles; Pratt, John; Seidenberg, A.
[1959]   "Intuitive Probabilities on Finite Sets," *Annals of Mathematical Statistics,* **30** (1959), 408-419.

Krantz, David H.
[1967]   "Extensive Measurement in Semiorders," *Philosophy of Science,* **34** (1967), 348-362.

Krantz, David H.; Luce, R. Duncan; Suppes, Patrick; Tversky, Amos
[1971]   *Foundations of Measurement,* Vol. I, Academic Press, New York and London, 1971.

Kuipers, Theo A. P.
[1978]   *Studies in Inductive Probability and Rational Expectation,* Reidel Publishing Co., Dordrecht, 1978.

Kyburg, Henry E., Jr.
[1961]   *Probability and the Logic of Rational Belief,* Wesleyan University Press, Middletown, 1961.
[1968]   "Bets and Beliefs," *American Philosophical Quarterly,* **5** (1968), 54-63.
[1970]   *Probability and Inductive Logic,* The Macmillan Co., Collier-Macmillan Limited, London, 1970.
[1974]   *The Logical Foundations of Statistical Inference,* Reidel Publishing Co., Dordrecht and Boston, 1974.
[1976]   "Chance," *Journal of Philosophical Logic,* **5** (1976), 355-393.
[1977]   "Randomness and the Right Reference Class," *Journal of Philosophy,* **74** (1977), 501-520.

Kyburg, Henry E., Jr.
[1978a] "Subjective Probability: Considerations, Reflections, and Problems," *Journal of Philosophical Logic,* 7 (1978), 157-180.
[1978b] "Review of Hacking: The Emergence of Probability," *Theory and Decision,* 9 (1978), 205-217.

Kyburg, Henry E., Jr. and Nagel, Ernest
[1963] *Induction: Some Current Issues* (ed. Kyburg and Nagel), Wesleyan University Press, Middletown, 1963.

Kyburg, Henry E., Jr. and Smokler, Howard E.
[1964] *Studies in Subjective Probability,* First edition (ed. Kyburg and Smokler), Wiley and Sons, New York, 1964.

Laplace, Pierre Simon de
[1825] *Essai philosophique sur les probabilités,* Fifth Edition, Bachelier, Paris, 1825.
[1952] *A Philosophical Essay on Probabilities* (English Translation of Second Edition of [L1825] by Truscott and Emory), Dover, New York, 1952.

Lehman, R. Sherman
[1955] "On Confirmation and Rational Betting," *Journal of Symbolic Logic,* 20 (1955), 251-261.

Lehmann, E. L.: *See* Hodges, J. L. and Lehmann, E. L.

Lehrer, Keith
[1969] "Induction: A Consistent Gamble," *Nous,* 3 (1969), 285-297.

Levi, Isaac
[1961] "Decision Theory and Confirmation," *Journal of Philosophy,* 58 (1961), 614-625.
[1964] "Belief and Action," *The Monist,* 48 (1964), 306-316.
[1966] Review of I. J. Good, *The Estimation of Probabilities*; I. Hacking, *The Logic of Statistical Inference,*; and R. C. Jeffrey, *The Logic of Decision, Synthese,* 16 (1966), 234-244.
[1967] *Gambling with the Truth: An Essay on Induction and the Aims of Science,* Alfred A. Knopf, New York, 1967. (Reprinted by M.I.T. Press, Cambridge, 1973.)
[1968] "Probability Kinematics," *British Journal for the Philosophy of Science,* 18 (1967-68), 197-209.
[1970] "Probability and Evidence," in *Induction, Acceptance, and Rational Belief,* ed. Swain, Marshall, Reidel Publishing Co., Dordrecht, 1970, 134-156.
[1974] "On Indeterminate Probabilities," *Journal of Philosophy,* 71 (1974), 391-418.
[1977] "Direct Inference," *Journal of Philosophy,* 74 (1977), 5-29.
[1978] "Coherence, Regularity and Conditional Probability," *Theory and Decision,* 9 (1978), 1-15.

Lewis, David K.
[1976] "Probabilities of Conditionals and Conditional Probabilities," *Philosphical Review,* 85 (1976), 297-315.

Lindley, Dennis V.
[1953] "Statistical Inference," *Journal of the Royal Statistical Society (Series B),* 15 (1953), 30-76.
[1958] "Professor Hogben's Crisis—A Survey on the Foundations of Statistics," *Applied Statistics,* 7 (1958), 186-198.

Lindley, Dennis V.
[1961]   "The Use of Prior Probability Distributions in Statistical Inference and Decision," Vol. 1 of *Proceedings of the Fourth Berkeley Symposium on Mathematical Statistics and Probability* (ed. Jerzy Neyman), University of California Press, Berkeley, 1961, 453-468.
[1965a]   *Introduction to Probability and Statistics Part 1. Probability,* Cambridge University Press, Great Britain, 1965.
[1965b]   *Introduction to Probability and Statistics Part 2. Inference,* Cambridge University Press, Great Britain, 1965.
[1971]   *Making Decisions,* Wiley and Sons, New York, 1971.
[1972]   *Bayesian Statistics, A Review,* SIAM, Philadelphia, 1972.
[1975]   "The Future of Statistics—A Bayesian 21st Century," *Supplement of Advanced Applied Probability,* 7 (1975), 106-115.

Lindman, Harold: *See* Edwards, Ward; Lindman, Harold; Savage, Leonard J.

Locke, John
[1965]   *Essay Concerning Human Understanding,* Collier Books, New York, 1965.

Luce, R. Duncan
[1959a]   *Individual Choice Behavior: A Theoretical Analysis,* Wiley, New York, 1959.
[1959b]   "A Probabilistic Theory of Utility and its Relationship to Fechnerian Scaling," *Measurement* (ed. Churchman and Ratoosh), 1959, 144-159.

Luce, R. Duncan; Krantz, David; Suppes, Patrick; Tversky, Amos: *See* Krantz, David; Luce, R. Duncan; Suppes, Patrick; Tversky, Amos.

Luce, R. Duncan and Raiffa, Howard
[1957]   *Games and Decisions: Introduction and Critical Survey,* Wiley and Sons, New York and London, 1957.

Luce, R. Duncan and Suppes, Patrick
[1965]   "Preference, Utility, and Subjective Probability," in *Handbook of Mathematical Psychology,* 3 vols., eds. Galanter, Luce, and Suppes, Wiley and Sons, New York, 1965, III, 249-410.

Mays, W.
[1963]   "Probability Models and Thought and Learning Processes," *Synthese,* 15 (1963), 204-221.

Marschak, Jacob
[1975]   "Personal Probabilities of Probabilities," *Theory and Decision,* 6 (1975), 121-153.

Menges, G.
[1970]   "On Subjective Probability and Related Problems," *Theory and Decision,* 1 (1970), 44-60.

Messick, Samuel and Gulliksen, Harold: *See* Gulliksen, Harold and Messick, Samuel

Mises, Richard von
[1951]   *Probability, Statistics and Truth* (Second Revised English Edition prepared by H. Geringer), Macmillan, New York, 1957.

Molina, Edward C.
[1931]   "Bayes' Theorem: An Expository Presentation," *Annals of Mathematical Statistics,* 2 (1931), 23-37.

Mosteller, Frederick and Bush, R.: *See* Bush, R. and Mosteller, Frederick

Mosteller, Frederick and Wallace, David
[1964]   *Inference and Disputed Authorship: The Federalist,* Addison-Wesley Publish-
          ing Co., Reading, 1964.

Nagel, Ernest
[1937]   *Principles of the Theory of Probability,* Vol. 1, no. 6 of International Encyclo-
          pedia of Unified Science, University of Chicago Press, Chicago, 1939.

Nagel, Ernest; Suppes, Patrick; Tarski, Albert (eds.)
[1962]   *Logic, Methodology and Philosophy of Science,* Stanford University Press,
          Stanford, 1962.

Nagel, Ernest: *See* Kyburg, Henry E., Jr. and Nagel, Ernest.

Neyman, Jerzy
[1947]   "Raisonment inductif ou comportement inductif?" Vol. 3 of the *Proceedings
          of the 25th Session of the International Statistical Institute,* Washington, D.C.,
          1947, 423-431.
[1952]   *Lectures and Conferences on Mathematical Statistics and Probability,* Second
          Rev. Ed., Department of Agriculture, Washington, D.C., 1952.
[1955]   "The Problem of Inductive Inference," *Communications on Pure and Applied
          Mathematics,* 8 (1955), 13-45.
[1957]   " 'Inductive Behavior' as a Basic Concept of Philosophy of Science," *Review
          of the International Statistical Institute,* 25 (1957), 7-22.

Neyman, Jerzy and Person, E. S.
[1932]   "The Testing of Statistical Hypotheses in Relation to Probabilities a priori,"
          Proceedings of the Cambridge Philosophical Society, 29 (1932-1933), 492-
          510.

Niiniluoto, Ilkka
[1972]   "Inductive Systematization: Definition and a Critical Survey," *Synthese,* 25
          (1972-73), 25-81.

Ore, Oystein
[1960]   "Pascal and the Invention of Probability Theory," *American Mathematical
          Monthly,* 67 (1960), 409-418.

Popper, Karl R.
[1955]   "Two Autonomous Axiom Systems for the Calculus of Probabilities," *British
          Journal for the Philosophy of Science,* 6 (1955-56), 51-57.
[1957]   "Probability Magic, or Knowing Out of Ignorance," *Dialectica,* 11 (1957),
          354-373.
[1959a]   *The Logic of Scientific Discovery,* Hutchinson, London, 1959. (Reprinted:
          Science Editions, New York, 1961.)
[1959b]   "The Propensity Interpretation of Probability," *British Journal for the
          Philosophy of Science,* 10 (1959-60), 25-42.

Pratt, John W.: *See* Kraft, Charles; Pratt, John W.; Seidenberg, A.

Pratt, John W.; Raiffa, Howard; Schlaifer, Robert
[1964]   "The Foundations of Decision under Uncertainty: an elementary exposition,"
          Journal of the American Statistical Association, 59 (1964), 353-375.
[1965]   *Introduction to Statistical Decision Theory,* McGraw-Hill Book Co., New
          York, St. Louis, San Francisco, Toronto, London, Sydney, 1964.

Raiffa, Howard and Schlaifer, Robert
[1961]   *Applied Statistical Decision Theory,* Harvard University Press, Boston, 1961.

Ramsey, Frank P.
[1950]   *The Foundations of Mathematics and Other Logical Essays,* (ed. Braithwaithe), Humanities Press, New York, 1950.

Ratoosh, P. and Churchman, C. W.: *See* Churchman, C. W. and Ratoosh, P.

Reichenbach, Hans
[1949]   *The Theory of Probability,* University of California Press, Berkeley, 1949.

Renyi, Alfred
[1955]   "On a New Axiomatic Theory of Probability," *Acta Math. Acad. Sci. Hungaricae,* 6 (1955), 285-335
[1970]   *Foundations of Probability,* Holden-Day, San Francisco, 1970.

Rescher, Nicholas
[1961]   "On the Probability of Non-recurring Events," *Current Issues in the Philosophy of Science* (Feigl and Maxwell, eds.), Holt, Rinehart, Winston, New York, 1961, 228-237.
[1966]   "Notes on Preference, Utility, and Cost," *Synthese,* 16 (1966), 332-343.

Robbins, Herbert
[1963]   "A New Approach to a Classical Decision Problem," *Induction: Some Current Issues* (ed. Kyburg and Nagel), Wesleyan University Press, Middletown, 1963, 101-110.

Rosenblatt, J. and Blum, J. R.: *See* Blum, J. R. and Rosenblatt, J.

Rosenkrantz, Roger
[1977]   *Inference, Method, & Decision,* Reidel Publishing Co., Dordrecht, 1977.

Rosett, Richard N.
[1965]   "Gambling and Rationality," *Journal of Political Economy,* LXXIII, No. 6 (1965), 595-607.

Ryll-Nardzewski, C.
[1957]   "On Stationary Sequences of Random Variables and the de Finetti's Equivalences," *Colloquium Mathematicum,* 4 (1957), 149-156.

Salmon, Wesley
[1963]   "On Vindicating Induction," *Induction: Some Current Isuses,* (ed. Kyburg and Nagel), Wesleyan University Press, Middletown, 1963, 27-41.
[1967]   *The Foundations of Scientific Inference,* University of Pittsburgh Press, Pittsburgh, 1967.

Savage, Leonard J.
[1951]   "The Theory of Statistical Decisions," *Journal of the American Statistical Association,* 46 (1951), 55-67.
[1954]   *The Foundations of Statistics,* John Wiley and Sons, New York, 1954.
[1959]   "La probabilità soggetivo nei problemi della statistica," *Induzione e statistica,* Istituto Matematico dell'Università, Cremonese, Rome, 1959, 75 pp.
[1960]   "Recent Tendencies in the Foundations of Statistics," *Proceedings of the International Congress of Mathematicians,* 14-21 August 1958, Cambridge University Press, Cambridge, 1960, 540-544.

Savage, Leonard J.
[1961a] "The Foundations of Statistics Reconsidered," *Proceedings of the Fourth [1960] Berkeley Symposium on Mathematics and Probability,* University of California Press, Berkeley, 1961, 575-585.
[1961b] "A Bibliography for the Foundations of Statistics and Probability," mimeo, 1961.
[1962a] "Subjective Probability and Statistical Practice," *The Foundations of Statistical Inference* (L. J. Savage and other contributors), Wiley and Sons, New York, 1962, 9-35.
[1962b] "Bayesian Statistics," *Recent Developments in Decision and Information Processes* (ed. Machol and Gray), Macmillan, New York, 1962, 161-194.
[1967] "Implications of Personal Probability for Induction," *Journal of Philosophy,* 54 (1967), 593-607.

Savage, Leonard J. et al.
[1962] *The Foundations of Statistical Inference,* John Wiley and Sons, New York, 1962.

Savage, Leonard J.: *See* Edwards, Ward; Lindman, Harold; Savage, Leonard J.
Finetti, Bruno de and Savage, Leonard J.
Friedman, Milton and Savage, Leonard J.
Hewitt, Edward and Savage, Leonard J.

Schick, Frederick
[1963] "Rationality and Consistency," *Journal of Philosophy,* 60 (1963), 5-19.
[1967] Review of Richard C. Jeffrey, *The Logic of Decision, Journal of Philosophy,* 64 (1967), 396-401.
[1970] "Three Logics of Belief," in Swain, ed., *Induction, Acceptance and Rational Belief,* 1970, 6-26.

Schlaifer, Robert
[1959] *Probability and Statistics for Business Decisions,* McGraw-Hill, New York, 1959.
[1961] *Introduction to Statistics for Business Decisions,* McGraw-Hill Publishing Co., New York, 1961.
See Raiffa, Howard and Schlaifer, Robert.

Scott, Dana and Krauss, Peter
[1966] "Assigning Probabilities to Logical Formulas," (ed. Hintikka and Suppes) *Aspects of Inductive Logic,* 1966, 219-264.

Seidenberg, A.: *See* Kraft, Charles; Pratt, John; Seidenberg, A.

Seidenfeld, Teddy
[1979] "Why I am not an objective Bayesian," *Theory & Decision,* forthcoming.

Shackle, G. L. S.
[1952] *Expectation in Economics,* Cambridge University Press, Cambridge, 2nd edition, 1952.

Shafer, Glen
[1976] *A Mathematical Theory of Evidence,* Princeton University Press, Princeton, 1976.
[1978] "Non-Additive Probabilities in the Work of Bernoulli and Lambert," *Archive for History of Exact Sciences,* 19 (1978), 309-370.

Shimony, Abner
[1955]   "Coherence and the Axioms of Confirmation," *Journal of Symbolic Logic,* **20** (1955), 1-28.
[1967]   "Amplifying Personal Probability Theory: Comments on L. J. Savage's 'Difficulties in the Theory of Personal Probability'," *Philosophy of Science,* **34** (1967), 326-332.
[1970]   "Scientific Inference," (ed. Colodny) *The Nature and Function of Scientific Theories,* University of Pittsburgh Press, Pittsburgh, 1970.

Siegel, Sidney: *See* Davidson, Donald; Suppes, Patrick; Siegel, Sidney.

Simon, Herbert
[1955]   "Prediction and Hindsight on Confirmatory Evidence," *Philosophy of Science,* **22** (1955), 227-230.

Skyrms, Brian
[1978]   "Statistical Law and Personal Propensities," *PSA 1978,* 2 vols., eds. Asquith, P. and Hacking, I., Philosophy of Science Association, East Lansing, 1979, v. 2.

Slovic, Paul; Fischoff, Baruch; Lichenstein, Sarah
[1977]   "Behavioral Decision Theory," *Annual Review of Psychology,* **28** (ed. Rosenzweig), Annual Reviews, Palo Alto, 1977, 1-39.

Smith, Cedric A. B.
[1961]   "Consistency in Statistical Inference and Decision," *Journal of the Royal Statistical Society (Series B),* **23** (1961), 1-37.
[1962]   "Comments on Savage," in *Foundations of Statistical Inference* (ed. Savage et al.), 1962, 58-62.
[1965]   "Personal Probability and Statistical Analysis," *The Journal of the Royal Statistical Society (Series A),* (General) **128,** Part 4 (1965), 469-499.

Smokler, Howard
[1965]   "Consistency and Rationality: A Comment," *Journal of Philosophy,* **62** (1965), 77-80.
[1967]   "The Equivalence Condition," *American Philosophical Quarterly,* **4** (1967), 300-307.
[1968]   "Conflicting Conceptions of Confirmation," *The Journal of Philosophy,* **65** (1968), 300-312.

Smokler, Howard E. and Kyburg, Henry E., Jr.: *See* Kyburg, Henry E., Jr. and Smokler, Howard E.

Sprott, D. H. and Godambe, V. P.: *See* Godambe, V. P. and Sprott, D. H.

Stigler, George
[1950]   "The Development of Utility Theory," *Journal of Political Economy, Part I,* **58** (1950), 307-327; *Part II,* **58** (1950), 373-396.

Spielman, Stephan
[1977]   "Physical Probability and Bayesian Statistics," *Synthese,* **36** (1977), 235-269.

Stegmuller, Wolfgang
[1973]   *Personelle und Statistische Wahrscheinlichkeit* (2 vol.), Springer Verlag, Berlin, Heidelberg, 1973.

Stelzer, John H.
[1967]   "Some Results Concerning Subjective Probability Structures with Semi-orders," Institute for Mathematical Studies in Social Sciences, Stanford University, Technical Report No. 115, Aug. 1, 1967, Psychology Series.

Suddith, William and Heath, David
[1976]   "On Finitely Additive Priors, Coherence and Extended Additivity," Technical Report 278 (November 1976), Department of Statistics, University of Minnesota.

Suddith, William and Purvis, Roger
[1976]   "Some Finitely Additive Probabilities," *Annals of Probability,* **4** (1976), 259-276.

Suppes, Patrick
[1965]   "The Role of Subjective Probability and Utility in Decision Making," Vol. 5 of *Proceedings of the Third Berkeley Symposium on Mathematical Statistics and Probability* (ed. J. Neyman), University of California Press, Berkeley, 1956, 61-73.
[1960]   "Some Open Problems in the Foundations of Subjective Probability," *Information and Decision Processes* (ed. R. Machol), McGraw-Hill, New York, 1960, 162-169.
[1974]   "The Measurement of Belief," *Journal of Royal Statistical Society, Series B* (Methodological), **36** (1974), 160-191.

Suppes, Patrick: *See* Davidson, Donald and Suppes, Patrick
          Davidson, Donald; Suppes, Patrick; Siegel, Sidney
          Nagel, Ernest; Suppes, Patrick; Tarski, Alfred

Tarski, Alfred: *See* Nagel, Ernest; Suppes, Patrick; Tarski, Alfred.

Teller, Paul
[1973]   "Conditionalization and Observation," *Synthese,* **26** (1973-74), 218-258.

Tintner, Gerhard
[1941]   "The Theory of Choice under Subjective Risk and Uncertainty," *Econometrica,* **9** (1941), 298-304.

Tversky, Amos
[1975]   "A Critique of Expected Utility Theory: Descriptive and Normative Considerations," *Erkenntnis,* **9** (1975), 163-173.

Tversky, Amos and Kahneman, D.: *See* Kahneman, D. and Tversky, Amos

Tversky, Amos; Krantz, David; Luce, R. Duncan; Suppes, Patrick: *See* Krantz, David; Luce, R. Duncan; Suppes, Patrick; Tversky, Amos

Van Frassen, Bas
[1976]   "Probabilities of Conditionals," in *Foundations of Probability Theory, Statistical Inference and Statistical Theories of Science,* 3 vols., Harper, William and Hooker, Clifford, eds., Reidel Publishing Co., Dordrecht, 1976, v. 1, 261-300.

Venn, John
[1886]   *The Logic of Chance,* Macmillan, London, 1888 (Reprinted, Chelsea, New York, 1963). '

Vickers, John M.
[1965]   "Some Remarks on Coherence and Subjective Probability," *Philosophy of Science,* **32** (1965), 32-38.
[1970]   "Probability and Non-Standard Logics," *Philosophical Problems in Logic,* (ed. Lambert), Reidel Publishing Co., Dordrecht, 1970, 102-120.
[1976]   *Belief and Probability,* Reidel Publishing Co., Dordrecht, 1976.

Villegas, C.
[1964]   "On qualitative probability σ-algebras," *Annals of Mathematical Statistics,* **35** (1964), 1787-1796.

Wald, Abraham
[1942]   *On the Principles of Statistical Inference,* Notre Dame University Press, South Bend, 1942.
[1950]   *Statistical Decision Functions,* Wiley and Sons, New York, 1950.

Williams, P. M.
[1976]   "Indeterminate Probabilities," in *Formal Methods in the Methodology of Empirical Sciences,* Przelecki, M., Szaniawski, R. and Wojeicki, R., eds., Reidel Publishing Co., Dordrecht, 1976, 229-246.
[1978]   "On a New Theory of Probability," *British Journal for the Philosophy of Science,* **29** (1978), 375-387.

Winsten, C. B.: *See* Barnard, G. A.; Jenkins, G. M.; Winsten, C. B.

Wolfowitz, J.
[1962]   "Bayesian Inference and Axioms of Consistent Decision," *Econometrica,* **30** (1962), 471-479.

Wright, George Henrik von
[1962]   "Remarks on Epistemology of Subjective Probability," *Logic, Methodology and Philosophy of Science* (ed. Nagel, Suppes, and Tarski), Stanford University Press, Stanford, 1962, 330-339.

Wrinch, Dorothy: *See* Jeffreys, Harold and Wrinch, Dorothy.

INDEX